Control Room Simulation

HUGH DAVID

Copyright © 2017 Hugh David

All rights reserved.

ISBN: 1979383820
ISBN-13: 978-1979383820

DEDICATION

M. Jean-Marc Garot (Leader and Liberator)

Mr. Charles Massie (Leader, Mentor and Friend)

Professor W. F. Floyd (Leader and Saviour)

Hugh David

Acknowledgements
(and Apologies)

Abbess Hildegard of Bingen
Miss Jane Austen
Mr. Anthony Hope
Sir Terry Pratchett
And others too litigious to mention

Hugh David

Contents

Dedication iii
Acknowledgements v
Contents vii
Figures viii
Chapter 1 – Introduction 1
Chapter 2 – Air Traffic Control 6
Chapter 3 – Types of Real-Time Control Room Simulation 10
Chapter 4 – Physical Structure 13
Chapter 5 – People 18
Chapter 6 – Defining a Simulation 22
Chapter 7 – Data 25
Chapter 8 – Experimental Design 30
Chapter 9 – Targets of Measurement 37
Chapter 10 – Recommended Measurements 41
Chapter 11 – Preparation 50
Chapter 12 – Running 64
Chapter 13 – Analysis 129
Chapter 14 – Reporting 156
Chapter 15 – Future Developments 167
Chapter 16 – Summary 174
Bibliography 177
Recommended Reading 180
Appendix 1 - Measurement Methods 181
Appendix 2 – Report 201
Appendix 3 – Conference Paper 276
Appendix 4 – Verbal Presentation 284
Appendix 5 - Poster 294
About The Author 296

Figures

Figure 1 Ruritania Airspace 26

Figure 2 Two Sector Split (1) 27

Figure 3 Two Sector Split (2) 27

Figure 4 - Initial Experimental Plan (1) 33

Figure 5 - Initial Experimental Plan (2) 35

Figure 6 - Massie Grid – Overall 47

Figure 7 - Massie Grid - TMA 48

Figure 8 - Control Room Schematic(1) 51

Figure 9 - Control Room Schematic (2) 52

Figure 10 - Final Experimental Plan 123

Figure 11 - Factor Coding 135

Figure 12 - Sector Position Factor 136

Figure 13 – Sectorisation Anovar (Main Effects) 137

Figure 14 – Duration Of Measure 138

Figure 15 - Run Means Anovar Main Effects 139

Figure 16 - Full Anovar - Sectors 140

Figure 17 Sectorisation Effect 141

Figure 18 - TMA Factors 142

Figure 19 -TMA Analysis Of Variance 144

Figure 20 - TMA Measures 145

Figure 21 – Overall Traffic 95% Weekday 152

Figure 22 - ISA Values 15th 0900 2 Sector W/E 100% Traffic 153

Figure 23 - NASA TLX components 154

Figure 24 - Contemporary Large Real-Time Simulator 168

Figure 25 - Contemporary Control Room 169

Figure 26 - Future On-Site Simulator 171

Figure 27 - Future General Purpose Simulator 173

Chapter 1 – Introduction

Real-Time simulation is an empirical technique for examining systems which include one or more human controllers. (It is sometimes called Man-in-the-loop simulation, to contrast it with purely computer simulation, in which the human element may be ignored or irrelevant. I consider that the term 'Man-in-the-loop' downplays the importance of the human element, suggests that the systems involved are amenable to mathematical analysis and is sexist to boot.) Most modern systems are computer-based, and simulation is a valuable tool for designing the human-computer interfaces needed.

To clear up a persistent ambiguity, the term "simulation" refers to the general process of simulating some activity. The term "a simulator" refers to the physical software and hardware used for simulation. The term "a simulation" refers to the process of using a simulator for a defined process. A "run", sometimes confusingly referred to as a "simulation" or a "simulation run", is a use of the simulator for usually one to two hours. An "organisation" refers to a set of "simulation runs" using a candidate method. A simulation may involve two or more organisations, each involving two or more runs.

In the modern context, simulation can be used in many industries, wherever a control room is being designed, staff trained and emergency drills developed. To name only a few, - aircraft, and air systems, trains and railway systems, ships and seaways, road tunnels, electricity generators and distribution networks, gas extraction and distribution systems, chemical plants and pipelines can all be simulated. Simulators cost money, but considered as means of avoiding costly mistakes and human tragedies they are extremely cost effective. Specific simulators will be needed to simulate particular vehicles, but systems for controlling many vehicles usually operate through a control room. Control rooms are increasingly designed around the human needs of the operators, and are becoming increasingly similar. For example, to simulate a train driver's cabin a very specific simulator is needed, while a railway signal control centre may look like any other control centre. It is often feasible to link one or more unit simulators into a network simulator, particularly where new types of units are being introduced into a network.

In Air Traffic Control Real-Time simulation is widely used for developing control rooms acceptable to the controllers. These are not necessarily the best solutions to future control problems, but they are generally accepted. In other contexts, for example in aircrew training, it is used primarily to accustom trainee aircrew to the physical 'feel' of the aircraft they will control, without the expensive and dangerous use of real aircraft. It may also be used to develop 'crew spirit' and accustom crews to working together – although this is less common in civil aviation than it should be. Military experience places considerable emphasis on developing this spirit, but many airlines now tend to treat aircrew as interchangeable units, regardless of the potential hazards of this approach.

Large Real-Time simulators are very expensive devices, and have generally been developed on an 'ad-hoc' basis by teams assembled for the purpose and dispersed on completion. There are good economic reasons for this, but it can lead to costly and embarrassing failures, particularly where the commissioners of the simulator are out of touch with the actual operating conditions. Given the speed of change in most technical domains, this occurs more often than one might expect. An Air Traffic Controller experienced with the traffic conditions of 2005 may be completely unaware of the pressures of 2015 while being sincerely convinced that he knows what the problems are or will be.

In the contemporary world, control rooms are increasingly designed to suit the physical and mental limitations of human operators, so that control rooms are increasingly similar in appearance and physical layout.

Although the physical specification of a simulator is relatively easy – often it may be simply a copy of an existing control room - the specification of the firmware, software, data, and 'wetware' is much harder, for reasons that will be discussed later.

The design and running of Real-Time Control Room simulators is essentially a craft – relying on empirical learning and experience rather than theory. This book is based on thirty years of experience in and around what was probably the most advanced and flexible Real-Time Control Room simulator in the world at the EUROCONTROL Experimental Centre.

This centre was founded in 1968, soon after the foundation of EUROCONTROL. EUROCONTROL (The European Organisation for the Safety of Air Navigation) is an international organisation. Its initial purpose was to provide upper airspace Air Traffic Control for the six founder members – The United Kingdom, Eire, France, Belgium, the Netherlands and West Germany. The EUROCONTROL Experimental Centre (EEC) was set up to develop the hardware and software for the Maastricht Upper Airspace Control Centre. It then went on to run simulations for the member countries of EUROCONTROL, taking advantage of the flexibility of its general-purpose digital Real-Time simulator – the first in the world.

As more countries joined EUROCONTROL (41 in all at the last count) and it became an advisory and coordinating body rather than an executive agency, the Experimental Centre diversified into many other areas of air traffic management. It retains its capacity for mounting Real-Time Control room simulations, using generic or adapted working positions and adapting its software to suit research and developmental studies.

Because the Experimental Centre is in the business of obtaining answers to specific questions, its orientation is rather different from that of a normal research institute. (On joining in 1970, I was told, "Here we do experiments. We don't do science!") Although there have been repeated attempts to achieve scientific respectability, practical considerations have always prevented this. This book refers to the "Craft" rather than the "Science" (or Art) of Real-Time Control room simulation.

The emphasis on obtaining specific answers, which are often translated into practice before any report is written, produced a general disinclination to publish in learned journals, or at all.

Military or civil-military simulations were usually restricted on general principles. Other simulations produced results that were not entirely welcome to the sponsor, who did not wish them to be circulated. Although the situation has since improved considerably, attendance at scientific conferences was originally regarded as an unearned holiday rather than a learning experience. Publication in any form was seen as an admission of under-employment. Neither activity improved promotion prospects. Consequently, very little of the accumulated experience was placed in the public domain.

For this reason, the references listed tend to be largely to EEC Reports or to my own publications during the years when I was one of the few people actually to appear in print. A further consideration is my old-fashioned reluctance to quote references that I have not read, a practice I recommend to readers of this book. If the result appears vain or even solipsistic, I can only apologise.

It is usually easier to understand new ideas if they come with concrete examples of what they mean in reality. I have concentrated on large scale Air Traffic Control simulators because I know something about these, although I have tried to extract general principles as far as possible. Small self-contained computer-based models – such as TRACON or TROTSKY, which involve one operator only, are better discussed separately. Although I have spent some years working with these, their problems and solutions are sufficiently different to confuse the general reader. Small simulators are only mentioned where they are needed to illustrate specific points.

To set the scene, Chapter 2 describes the Air Traffic Control system, as it exists in most developed countries. I have tried to avoid the torrent of acronyms that is the bane of Air Traffic Control discussion. Air Traffic management is usually a civil service environment, where success is founded on avoiding mistakes and never giving offence. The old Japanese proverb "It's the nail that sticks up that gets hammered" summarises the Air Traffic Control culture nicely. The maxim "When In doubt – reorganise" leads to the regular re-naming of elements of the Air Traffic Control system in the hope that this will substitute for the clear and rigorous thinking and unpopular decisions the system so desperately needs. The modern tendency to refer to Air Traffic Management rather than Air Traffic Control is a textbook example. In this book, I have tried to use clear non-technical language wherever possible in the full knowledge that this would blast any future prospects I might have had, if it had not done so long ago - but that is another story.

Chapter 3 discusses different types of Real-Time simulation, as used for training, exploring new methods and revising the allocation of airspace into sectors. It discusses the types of simulator appropriate for these different purposes.

Chapter 4 sketches what is involved in terms of investment in time, effort and resources (hardware and software) in building and running a large-scale Real-Time Control room simulator. Since my knowledge of finance, financial estimation and project management is about that of a gerbil, this is essentially a 'ball-park' estimate.

Chapters 5 to 14 describe the complete process of Real-Time Control room simulation, from the initial definition to the final report. To add interest, an example is provided. For the record, the characters (except one) and situations described are entirely fictitious. Any resonance with the real world is subject to the reader's perception.

First, the most important part of preparation, the people, staff and participants, are considered (Chapter 5). It is very important that these are participants – not subjects. They are usually trained Air traffic controllers, familiar with the area and its problems. As such they are a vital resource, and it is essential to make the best use of their knowledge.

Chapter 6 discusses the process of defining the aims of a simulation. This is the most important phase of any simulation, but is often skimped or taken for granted, leading to eventual disaster. (There have been occasions where the exercise of defining simulation aims has led to the discovery that there were not any, and that the simulation was pointless. Needless to say, these do not figure in the literature.)

Chapter 7 discusses the problems of obtaining the working material for simulations – the static data representing the physical environment and the dynamic data representing the traffic that is to be controlled.

A discussion of the thorny problems of experimental design follows in Chapter 8. Serious conflicts often occur between the requirements for scientific rigor and the practical interests of the participants. There are also problems in reconciling the serviceability problems of large-scale simulations and the balance of experimental design. (Science usually loses out.)

Chapter 9 discusses types of possible measurement, their targets and relevance to practical problems. Chapter 10 suggests measurements that are usually useful, and some that may be useful sometimes.

Chapter 11 gives examples of the preparation for a simulation, while Chapter 12 discusses the actual running of a simulation.

Chapter 13 gives a brief discussion on methods for the analysis of these measurements, concentrating on practical problems rather than on scientific methodology.

Chapter 14 gives a guide to the presentation of the results of a simulation – as a verbal presentation, a conference paper, and a technical report. These are three separate items, and require very different approaches. Different audiences require presentations adapted to their interests.

Chapter 15 suggests possible future developments.

Chapter 16, the final chapter, summarises the overall message of this book.

References and Recommended reading are also provided.

Appendix 1 is a description of some potential measurement methods.

Appendix 2 is a complete Real-Time Simulation Report, based on the simulation described in Chapters 5 to 14. Comparison of this report with those chapters may be educational.

Appendix 3 is a conference paper describing the same simulation.

Appendix 4 provides a 'PowerPoint' presentation of the simulation, to illustrate how the (fictitious) simulation would be presented verbally.

Appendix 5 shows a poster presentation of the same simulation.

Chapter 2 – Air Traffic Control

This is a generic description of the contemporary ATC system. It is not a catalogue of all possible or existing systems. It does not claim to be complete, but serves to set the scene for further discussion. Even within the current Eurocontrol area, there are regions having additional facilities, and others lacking some of the features described here.

The first element in the development and operation of a flight in which Air Traffic Control is concerned is Flow Management (originally called 'Flow Control' - a name now felt to have negative implications cf. Birth Control, Pest Control). It is not possible for the existing system to handle all the flights that wish to fly through the most densely occupied regions. This problem was initially tackled by each sector imposing limits to the number of aircraft that would be accepted on various routes, or from adjacent centres or sectors, on an ad-hoc basis. These individual, uncoordinated limits resulted in accumulating restrictions, and a consequent waste of capacity. One summer, a national centre (which cannot be named) situated at the crossing point of major European airways, found that it was receiving so much overflying traffic, which could not be significantly delayed because it was already in the air, that it could not get its own national traffic into the air at all (Godverdomme!). A centralised European Flow Management System is now in place, which allocates 'slots' to flights - specifying the time at which they may take off. This produced a significant reduction to the delays experienced by passengers, particularly during the peak (holiday) seasons. The steady growth of air traffic has swallowed up the improvement within a few years, so that few travellers are aware of it. The existence of Flow control is a tacit admission that the existing Air Traffic Control system cannot cope with demand.

The ICAO flight plan remains the cornerstone of day-to-day Air Traffic Control. Most standard commercial flights operate in accordance with repetitive flight plans agreed long in advance between the companies and the ATC authorities. Most flights use traditional flight plans filed a few hours before the aircraft departs, although the flight plans are computer-generated and presented electronically. The ICAO flight plan includes routes, specified as a series of beacons, and preferred flight levels. (The preferred flight level is in general the level most suited to the airline's economic priorities. ATC tries to provide that level, but may have to allocate another to avoid conflict with other aircraft.) Transatlantic flights are specified slightly differently, to take advantage of day-to-day shifts in the jet stream and other air movements.

Once a flight plan has been agreed, ATC knows almost all it needs about the flight, except that the actual time the aircraft leaves the ground is never really known in advance, even where Flow Management has allocated a slot.

The flight plan is then translated into 'strips'. In a surprisingly large number of centres, these are still physical strips of coloured card, with cabbalistic abbreviations printed – or even hand-written - on them. Usually, a strip is prepared for each flight in each sector, giving the planned details of the flight. Sometimes, a single strip is prepared for the passage through a centre, and handed, carried or thrown from sector to sector as needed.

Depending on the system, strips may be prepared beforehand on the basis of the assigned take-off time, generated electronically at the actual time of departure, or for more remote parts of the journey, produced at a defined time before sector entry (but after the aircraft is in the air). In the more advanced regions, strips are not used, although they may still be produced in the hope that controllers could revert to strip-based control if their radar systems failed. This procedure is akin to explaining to passengers how to don a life jacket in the event of an emergency landing on water.

The activities of the aircraft before it leaves the ground are not strictly the concern of ATC, although Eurocontrol has carried out studies to analyse the degree to which aircraft keep to their assigned 'slots' and to attempt predictions of actual take-off time based on earlier events. These studies tended to show that the actual time of take-off is often the first definite indication that the flight will take place, and is the first reliable indication of the actual time of the flight.

After departure, the flight usually follows a Standard Instrument Departure, which defines the route the aircraft will take from the runway until it leaves the Terminal Manoeuvring Area. This route is pre-defined, including restrictions on height, to reduce perceived noise at ground level, to avoid conflicts with arriving aircraft at other levels or to avoid rising ground or obstacles. The flight should arrive at the edge of the Terminal Manoeuvring Area at a predetermined height and position. (In practice it may be climbing, so that the uncertainty of its actual height may be considerable. Many pilots climb as fast as they can, regardless of instructions to maintain a particular rate of climb. Sometimes heavily loaded aircraft cannot climb at the rate required. In either case, some pilots do not inform ATC of their deviation, apparently hoping it will not be noticed.)

After leaving the Terminal Manoeuvring Area, the flight will pass through a succession of en-route sectors. A sector is usually defined as a volume of airspace defined by lines on a map and upper and lower flight levels, although airspace may be reserved within this volume for military purposes, for Terminal Manoeuvring Areas, or to protect vulnerable features on the ground. Traditionally, airways are defined by a sequence of beacons on the ground. In each sector a Planner will consider the planned flight profile, verifying that there are no conflicts with other aircraft, usually in terms of separation time and cruising flight level at selected beacons on the route. If there are conflicts between two or more aircraft, he will try to find a solution by changing the assigned flight level of one or more aircraft. This may involve co-ordination with the planning controller of the next sector, or a notification, by voice, nudge, strip manipulation, code mark or electronic tag to the Executive that he will have to change the flight level(s) assigned. In some cases, the Planner will not be able to find a conflict-free trajectory and will leave conflict resolution to his partner, the Executive.

In principle, the Planner should not need to intervene in a flight after it has entered the sector, although he must retain information on it to compare with other aircraft. In general the Planner compares future heights and times over fixed points. The Planner acts by changing plans for the future, and communicates these planned changes to the Executive and to neighbouring Planners.

The Executive acquires control of the aircraft when or before it enters the relevant volume of airspace The aircraft usually contacts him by radio about this time and verifies the actual position of the aircraft on his radar or System Data Display. (A System Data Display is a large, high-definition display on which not only the radar data but also the associated planning data, predicted conflicts and entry or exit profiles may be shown.) The Executive usually carries out a check for potential conflicts or for deviations from track. He may give changes of level required by the flight plan or for conflict avoidance at this time or later. The Executive maintains a watch on the flights in his area, and may order changes of direction to avoid conflicts. He may request speed changes, either directly as a number of knots or a Mach number (a fraction of the speed of sound), or as a time over an exit point. He may also instruct a flight to proceed direct to a later point, cutting off a corner, or to return to its flight-planned track. Eventually, at the boundary or before the flight leaves his airspace, he instructs it to change to the frequency of the next sector and hands it over to the next Executive. It is good practice, although not always possible, to hand it over at the correct flight level and on track. At present, although the technology is available, most controllers do not adjust aircrafts' speeds to ensure that the aircraft leave at the time originally planned. The difference is rarely more than a few minutes.

The definition of 'radar conflict' usually employed is a good example of an 'evolutionary fossil' similar to the vermiform appendix in human anatomy. Normally defined as five nautical miles linear separation between aircraft, this standard appears to derive from the size of the radar 'paints' on early radar displays. However, two modern airliners approaching head-on at full speed would be in an extremely perilous situation well before reaching five miles separation, while the same airliners proceeding on parallel or diverging tracks at a separation of three nautical miles would be quite safe. Controllers (understandably) and managers (unforgivably) cannot grasp that alternative definitions, possibly based on wake turbulence, may be safer and make better use of airspace.

The primary tool of the Executive is the Radar or System Data Display, although he may be obliged to refer to strips to determine flight intentions, and other details, and to observe any Planner modifications. (In actual practice, both Planner and Executive are sufficiently familiar with most of the traffic that they know that a certain call sign represents a certain flight plan, which will take place about a certain time, using a certain type of aircraft and requesting a certain flight level. If any of these expectations are changed, considerable additional effort may be required.) Essentially the Executive works in and around the present time. He acts by communicating with the aircraft, giving instructions that are (usually) for immediate action. ICAO standards lay down that the pilot should repeat back every instruction and that the Executive should listen to the repeat and acknowledge that it is correct. This is inconsistent with what we know about human cognitive mechanisms. It is physiologically impossible for human beings to maintain sustained attention to purely routine activities for long periods. In reality, the standards are rarely observed, because there simply is not time, because aircrew or controllers may not feel the need for them, because it is a national tradition to treat 'the frequency' as a social medium, or because they do not fit with changed circumstances. The main use of standards appears to be for fixing the blame for accidents on the unlucky pilot or controller who happens to be in charge when an accident happens. No-one blames the manager who yields to commercial pressure by cutting corners or the bureaucrat who finds it expedient to impose impossible working practices rather than assert the need for properly designed and adequately manned systems.

When approaching the destination airport the flight will normally enter a Terminal Manoeuvring Area and employ a Standard Arrival Route. In what are now less frequent instances, it will be directed into a 'stack' where it will fly a 'race-track' oval pattern, descending to lower flight levels as preceding aircraft land (and, if only one runway is in use, depart) until it is directed to land under Approach Control. Terminal Manoeuvring Areas can be extremely complex structures, involving interleaved Standard Instrument Departures and Arrival Routes to several airports and 'stacks'. Other features, such as intruding terrain or prohibited areas, may limit freedom to manoeuvre. Many of the aircraft in the Terminal Manoeuvring Area are climbing or descending. In some large Terminal Manoeuvring Areas, corridors for light aircraft further complicate the problems.

In practice, life is rarely as clear-cut as is implied above. The Executive and Planner work as a team, sharing tasks and checking each other's work. Where traffic is light, for example in parts of Canada, Executive and Planner may be the same person, an inherently dangerous procedure. Eye-movement studies have shown that Planners may spend up to 40% of their time looking at the radar. The Executive may discuss height allocations with the Planner. Non-verbal communication goes on between the Executive and Planner. For example, the planner may displace a strip in its holder to warn the executive of a change. These deviations from the schema are typical of real systems, and exemplify the adaptability of the human operator, who will find a way of making even the most clumsy system work. (In fact, "working to rule" will usually cripple any major co-operative system.)

The current system has 'evolved'. (This is not a compliment.) As tools became available, they were brought in and added to the existing system. At no time did anyone stop and think about the system as a whole. The various improvements have steadily increased capacity, but at the same time, it has been necessary to reduce sector sizes, and multiply coordination and supervision processes, and to place more and more pressure on the controllers. Eventually (sooner rather than later) a limit will be reached. This will probably take the form of increased delays at airports, the use of legal or illegal stimulants by controllers, increased sickness among controllers, and increased incidence of electronic system failures rather than an increased incidence of accidents. Some readers may recognise these symptoms in other areas of study.

Chapter 3 – Types of Real-Time Control Room Simulation

Simulators are widely used in training, for individuals and small groups. Simple simulators may be cost-effective in early training, to assist in implanting 'drills' in long-term memory. More elaborate and realistic simulators appear to be equally cost-effective in conversion training of, for example, airline pilots. As Dubey (2000) points out, these simulators are used to instil attitudes and form emotionally bonded working groups. Any measurement of performance is incidental.

Training

'Training' simulators and simulations have shown themselves effective in aircrew training, and in basic training in aviation and Air Traffic Control. They allow trainees to familiarise themselves with the 'nuts and bolts' of equipment, and to practice pre-defined routines and drills that rarely occur in reality. They can also speed the process of team formation. Practically, training simulators are more or less permanent installations, great care being taken to make them resemble the aircraft or control positions for which the participants are being trained.

(Aer Lingus, the Irish national airline, named its aircraft after famous Irish saints – St. Patrick, St. Brendan and so on. Their flight simulator was named 'St. Thetic'.) The flight characteristics of the aircraft are very carefully simulated, and elaborate traffic situations are developed. This is cost effective because the same simulations can be re-run repeatedly with different groups of trainees.

A Training Simulator will be used by skilled instructors, who will usually be experienced operators themselves, with the extra knowledge and experience to control the running of the simulation, triggering simulated malfunctions, or introducing unexpected weather conditions and otherwise stressing the trainees, or skilled staff 'converting' to a new model or a new type of aircraft. They may sometimes be assisted by operators playing the part of other aircraft, air traffic control or other participants.

The book 'Flight Simulation' by Lee (2005) provides a very complete guide to the use of Real-Time Control room Simulation in training.

Organisational

Most large Real-Time Air Traffic Control room simulations are 'organisational' used to 'validate' proposed changes in the sector layout for a volume of airspace, which may include parts of several countries, and the interfaces between national ATC systems. These simulations are usually organised on a strictly pragmatic basis, and usually involve comparisons with as close a simulation as practical of the current system. It is rarely economic or practical to provide as exact a replica of the real world as is done in training simulators. A large Real-Time Control room simulation has been described as 'A psycho-drama trying to be a scientific experiment.' Changes in the course of the running of the simulation are often made, so that a compromise between two possible organisations can be tried out and unpromising alternatives abandoned during the later part of the simulation. This leads, inevitably, to what statisticians call 'confounding', since the effects of different organisations cannot be distinguished from learning effects. Since the main intention of organisational simulations is to convince the controllers that the proposed changes are practical, the discomfort of statisticians is acceptable, except perhaps to the statisticians. The opinions of the controllers are the main determinant of success in these simulations, and 'scientific' measurement is not highly regarded.

Developmental

A closely related type of simulation is the 'developmental' simulation. This maintains the traffic and organisation of a specific area and investigates the effects of changes is the equipment or methods employed. These changes might involve for example, using one Planner and two Executives in a sector, or the use of a different type of display. Usually, an existing area of airspace is used, with traffic derived from that usual in the area, and with participating controllers who are familiar with the area. Once again, there will be problems of familiarisation, particularly when the current system is run at the start of the simulation as a training and familiarisation exercise. Here again, it is the opinions of the participating controllers which determine the acceptability of the proposed changes. 'Scientific' or 'objective' measurements are often attempted in this type of simulation, although few such simulations would be accepted for academic theses

Research

The fourth type of simulation is the 'research' simulation. Usually, this is intended to study an alternative method or an imported tool which may improve the control process. Here again, an existing airspace is used, with, usually, staff controllers from the simulation centre. Scientific respectability is sought by attempting to apply the experimental psychology paradigm. This paradigm, in essence, assumes that when a few variables are changed under control, the changes in measured variables are due to these changes. There may be some uncontrolled 'nuisance' variables, which can be accounted for by proper experimental procedure, but the underlying situation does not vary systematically during the experiment. In ATC, these assumptions simply do not hold. The results obtained are rarely sufficiently well controlled to achieve statistical significance, and many 'nuisance' variables are simply unknown or ignored.

(This is not to say that large Real-Time Control room simulators have no role to play in the development of ATC. They are invaluable as the final validation stage, where controllers must be convinced that a system change is acceptable.)

Other, more cost-effective, methods are available for research in large-scale systems. Traditional ergonomics methods may be used in equipment design. Human-computer interface design methods are now well established.

Fast-Time

Sophisticated 'fast-time' models may be used to assess different work organisations - provided that sufficiently reliable performance data is available. A simple 'fast-time' model can be useful in checking that samples of traffic do not contain significant anomalies. Such a model can be used to ensure that there are no 'traffic jams' where aircraft enter the system.

Fast-time simulations can be misleading however, as a comparison of a Real-Time control room exercise and a fast-time simulation of the same traffic and area showed. The fast-time system always changed the flight path of the aircraft entering the area to solve the immediate conflict rather than finding a solution to the set of conflicts arising, so that a single aircraft entering at a wrong flight level could lead to several succeeding aircraft being given changed flight levels. Equally, where several air routes converged just outside the simulated area, the fast-time model would deliver several aircraft at the same level, giving the next sector an impossible task. Real controllers instinctively separated the departing aircraft.

Observation of real ATC, given an understanding of what is actually going on, is a useful check on theoretical speculation - for example, conflict detection and resolution, subjects of much theoretical interest, rarely actually occur in current practice, and where they do, are not separate activities. In 'real life' controllers act to 'stream' traffic so as to prevent conflict situations occurring. In a manner not unique to ATC, controllers tend to assume that experimenters know what they are doing and, consciously or unconsciously, obtain cues from the questions they are asked, or the way in which they are asked.

Chapter 4 – Physical Structure

Hardware

A Real-Time Control room simulator is, as far as its physical appearance goes, a control room. Real-Time Control room ATC simulations have involved up to 90 participants at individual working positions. These working positions were traditionally designed to accommodate racks of 'Strips' for the Planner, or a Radar display for the Executive. Modern systems use large computer displays – often 2000 x 2000 pixels on a 50cm. square screen. These displays, weighing about 50 kilograms, required special supports, usually wheeled and cantilevered over the desk provided so that they could be easily moved and replaced. They were originally specified to display a digital copy of the traditional nineteen inch circular radar display, but are now used as System Data Displays, where different aspects of the stored flight data and the data collected from on-line systems – radar processors or GPS data transmitted by digital data link are displayed in different ways in different windows. They are now being replaced by flat-panel displays in many modern centres. There is, however, a regrettable reluctance in Air Traffic Control organisations to acquire modern equipment. The saying goes that:

"In Air Traffic Control,

- They do not replace equipment when newer equipment becomes available.
- They do not replace equipment when the manufacturers stop supplying spares.
- They do not replace equipment when there is none left on the second hand market.
- They start planning replacement when the scrap merchants have nothing left – if then."

The choice of displays and their location on the display surface differs for Planners and Executives, and may vary between sectors. Controls for traditional Radio Telephony - frequency selectors, volume controls and press-to-talk switches - may also be incorporated, or may be provided separately according to the requirements of the simulation. The working position will now usually have a point-and-click device, such as a mouse or track ball, and may have a keyboard, although controllers are not usually expected to do much typing. (In accord with the last paragraph, there are no touch-screen interfaces in use in ATC, although there have been some studies of specialised applications.) ATC track balls are traditionally large, about the size of a tennis ball, which seems to provide quicker centring on screen. Current commercial standard trackballs are about golf ball size, and are not liked by controllers. The author has been involved in interface design, and on one occasion had the delicate task of informing the Royal Air Force that their balls were too small. The overall layout is not usually changed and windows are not overlaid during control work, although lists of data may vary in size and controllers may open windows associated with aircraft to see coded flight plans or enter changes to flight plans. (Every second spent in manoeuvring windows is taken from the controller's proper activity – and in early versions, controllers occasionally 'lost' vital windows by accidentally minimising them and being unable to find the minimised icon. When an

Executive loses the radar screen in this way, he rapidly becomes distressed.) Worse, the system might 'pop-up' a warning message just where the controller's attention was directed. A more subtle problem is 'falling into the screen', where the controller's attention is concentrated on a problem in one corner of a large display, so that he fails to notice a problem developing elsewhere.

Working positions are usually arranged in pairs, for a Planner and an Executive who control a sector, and known as a Sector Suite. (Assistant Planners, Assistant Executives, or just plain Assistants may sometimes be found, usually in less developed centres.) These sector suites are usually arranged in roughly the same way as the airspace they control. Where two sectors control different ranges of height for the same airspace, this simple disposition does not work.

In many modern control rooms, one single very large display screen is also provided, giving an overall view of the system being simulated. There is some dispute about the economic value of such screens, but it can be argued that they provide an overall view of what is currently going on, which may be valuable for strategic planning. Cynics suggest that their main use is to impress visitors.

In addition to the Sector suites for the sectors being simulated, a number of working positions are provided to simulate the adjacent sectors, TMAs and so on. These positions are used by 'ghost' or 'feed' controllers. The description 'dummy controllers' is not appreciated. The controllers at these positions hand over inbound aircraft and receive outbound aircraft, making sure they are properly separated and normal handover procedures are followed. These positions may be manned by controllers from the region being simulated or by controllers permanently employed at the simulation centre. Their work is rather different from that of normal controllers, since they do not have to observe the normal constraints for traffic within their sectors, provided they do not disturb the image provided for the simulated area. These working positions are usually placed in the same room as the simulated sector suites.

There will usually be two or three 'Supervisor' positions. Originally these were placed in a room between the 'pilots' and the participant controllers. However, it has become accepted over years of experience that these positions are better placed in the simulated control room, preferably raised so that they have an overview of the participants. One of these positions is occupied by the simulation supervisor. This is usually the project leader. This position provides on-line information about the progress of the simulation (including the ISA reports – see Chapter 12) and has facilities for copying controllers' displays. It may have other facilities as necessary, such as CCTV displays from strategic points of the control room. A second position, usually used by the technical supervisor, provides the controls needed to start and stop the simulator, and its recording equipment.

Simulators usually have "pause" and "re-start" facilities, although these are usually more trouble than they are worth. While computers can be paused, human beings cannot, nor can they be re-set. The behaviour of controllers before and after a pause is usually rather different from what would be expected if the simulator ran normally. Although re-starts may be tolerable when the simulator is just starting a run, they are undesirable if the traffic has been shown to the controllers, since it will become familiar and problems may be anticipated in an unrealistic fashion.

Facilities for visitors, if unavoidable, should not allow visitors to interfere with the activities in the simulation room. It is desirable to have a viewing gallery, preferably with a glass barrier between it and the working simulation room. Here is where the large screen comes into play. Chapter 11 includes advice on the delicate topic of how visitors should be handled.

An adjacent room contains a similar number of working positions for the 'pseudo-pilots', who simulate the pilots of the aircraft. These 'pseudo-pilots', who are not usually qualified aircraft pilots, speak 'radio' voice messages on prompting by the computer controlling the system, and input standardised messages corresponding to the instructions given by the controller to the simulated aircraft. One or, sometimes, two 'pseudo-pilots' will usually handle communications for several aircraft in a particular sector. Different pilots will handle the aircraft in previous and subsequent sectors. It is too costly to employ one pilot per aircraft simulated, but the use of one or two 'pseudo-pilots' per sector has several disadvantages. In 'real life' pilots have different voices and aircraft have different background noises, which provide identity clues to the controllers. Real pilots do not observe the formal rules for speech communication as strictly as they should. (They do not always state their call-sign in each message of a conversation, and do not read back every message in full. There is a 'social' element in communication which may not be formally recognised, but plays a real part in assuring pilots and controllers that their opposite numbers are awake and aware of them (Mell, 1992) . 'Pseudo-pilots' are usually too well trained, and do not make the sorts of mistakes that real pilots do. Attempts have been made to develop computer simulations of pilots, using automatic speech recognition systems, but these are generally useful only in training, where rigid speech formulations may be preferable and prolonged training of the speech recognition system is acceptable. Experienced controllers tend to employ more relaxed and idiomatic speech patterns than are officially specified.[1] Voice distortion software and synthetic background noise are technically feasible to help differentiate aircraft within a sector, but have not, as far as I am aware, been introduced in any large simulators.

A final, apparently trivial, element, of a Real-Time Control room simulator is a Participants' Room. I hesitate to use the term 'rest room' as that phrase like 'bath room' has different meanings on either side of the Atlantic, with potentially disastrous but occasionally Rabelaisian implications. This should provide sufficient space and facilities for participants to relax between simulation runs, and to store coats and so on during runs. It should also, by providing tea, coffee, biscuits and participants' local newspapers, send a message that they are valued and respected collaborators. Non-participants should be admitted only by invitation of the participants.

Finally, the location of the simulator itself is important. Although a training centre may have its own stand-alone simulators, it is probably better to locate a simulator at a working control centre. Ideally, it should be located close to the real control room. In future systems the emphasis will be shifting from heavy routine activity to 'control by exception' in which

---

[1] "Air France 513, where the hell are you going, scheisskopf?" (Personal Observation)

controllers monitor traffic, intervening only when changes to planned flights are needed. In order to maintain controllers' skills, a significant proportion of controllers' working shifts would need to be devoted to simulator exercises in a situation as close to the real environment as possible.

Software

The software structure of a large Real-Time Control room simulator is necessarily complex. I am not in a position to provide much useful information about the design architecture of large simulators, although I can offer some pointers based on practical experience.

The core of a Real-Time Control room ATC simulator is the system that simulates the physical movement of the aircraft involved. This may produce an updated image at pre-set intervals, simulating older radar-based systems, or newer systems which themselves simulate a radar system, although they usually update the full screen rather than simulate a rotating scan. Alternatively it may employ an asynchronous system, generating events from prepared files, or by processing events. This latter system is more flexible but much more complicated.

In recent years, there has been a steady tendency for control rooms to become more alike, and this has been reflected by a tendency for the software to be constructed from modules that may be common to a field of simulation, such as ATC, Rail, Road or Shipping traffic, or pipeline, power distribution, water or waste water distribution. In addition, analysis and replay modules may be generally applicable

In any case, events should be marked with the time at which they should occur. One non-professional simulation was prepared and used relying on the speed of the system to approximate to real time. Unfortunately, when transferred from a Russian PC to a US version, the entire thirty-minute simulation took exactly eight seconds ... A quick-thinking programmer sprinkled the system with dummy loops counting to a thousand, slowing it up sufficiently to be visible. Great care must be taken when 'simultaneous' events occur. In one 'fast-time' simulator (a mathematical model where human operators are not involved) events occurred at nominal six-second intervals. Traffic was programmed to leave the simulation at beacons around the area of interest. It was also programmed to carry out various actions at each beacon (changing course, etc.) Unfortunately, sometimes the 'leave the sector' event came before the 'pass the beacon' event. Chaos ensued. We eventually discovered that system was clearing away the aircraft and closing all relevant files – then trying to simulate the passage of the beacon by a now non-existent aircraft.

Whether synchronous or not, a clue to imminent failure is often available when the 'simulator time' begins to fall behind real clock time. There is not usually much you can do about it, but it impresses management, and infuriates system operators, if you can announce "The system is about to fail". (It is for you to decide which outcome is more important to you.)

It is not usually possible to carry out all aircraft generation, display and recording on one machine. Although traffic data and fixed environmental data can be prepared beforehand, once the operators begin to interact with the system simulations invariably diverge, so that it is practically impossible to repeat any simulation exactly (Hugh David, 2004). In the case studied, the same scenario was repeated 52 times, the first divergence taking place after three minutes. After 30 minutes, all runs were different.

Almost all large simulators are modular, with separate software packages - and sometimes separate hardware packages - controlling, for example, the generation of traffic from the prepared flight plans, the display of the current situation, the dialogues between the controllers and the simulated aircraft, the dialogues between the 'simulator 'pilots' and the system and the recordings of all these processes.

From time to time, it will be necessary to modify or update some of these modules because techniques or technology have changed. The exact definitions of the interfaces will rarely be available, and the original designers will equally be unavailable, through promotion, transfer, end of contract, retirement or even death. It is therefore of paramount importance to maintain the existing system in working order until the revised system has been developed and extensively tested. It is also very unwise to replace two large modules at the same time, especially if different teams or subcontractors are employed.

Ideally, a simulator should use the same software as the system it simulates. Unfortunately, this system may not yet exist, may be itself flawed and being 'updated' or may use proprietary software which cannot be modified. Ideally, all the simulator need do is substitute synthetic inputs for those coming from the real world.

It should, in theory, be possible to use the system recordings made by the real system to record the activity of real controllers. In practice, more or different information is needed from a simulator, and separate recording methods will be needed. Simulator software usually has to be written by a small team in a relatively short time and cannot aspire to the reliability that is required of operational systems. (For example, real systems must allow for the discontinuity of time around midnight. Simulators usually can ignore this.)

The design, building and testing of highly reliable safe dynamic systems is a slow, expensive task for specialised professionals. The synchronisation of recordings is very important. Where data are recorded in packets, small delays due to the data collection and compression algorithms may produce wildly distorted results.

Chapter 5 – People

A concert hall, however luxurious, arrays of instruments, however shiny and expensive, and musical scores, by the score, do not make an orchestra. The most important components of an orchestra are the people who play the instruments, and the conductor, who directs their efforts. (Real musicians would include the audience, but that is another story.) Similarly, buildings, hardware and software may be the visible components of a real time simulator, but the most important parts of a Real-Time simulation are the people who carry out the simulation. These people can be divided into two groups, the staff, who are more or less permanent, and the participants, who carry out the simulation and usually are practicing operators with experience of the area.

Staff

In any Real-Time Control room simulation, the key figure is the simulation leader. Experience has shown that simulation leaders must be experienced in the field being simulated. They must also be familiar with the simulator, its limitations and possibilities. They should also be sufficiently literate to write a report on the simulation and sufficiently articulate to present the conclusions in an acceptable form. It is therefore usually unwise to bring in a leader from the region being simulated, since he or she will be distracted by learning the ropes while steering the ship. It happens that we have an experienced controller, once an RAF officer, an amateur pilot with a wide and deep experience of how people think in aviation. We will call her Elizabeth (Liz) Bennett.

The next most important person is the assistant simulation leader. He or she must complement the simulation leader, and carry out the tasks that she has not time for. It is usual to use this position as a testing ground for potential simulation leaders, since the difficulties arising from poor leadership are daunting, but it is virtually impossible to judge the potential of candidates without extensive observation. (Where it becomes clear that an assistant simulation leader is not up to the top job, he is usually discretely channelled into a less demanding job. At least, this is what happens in civilised countries. Across the Atlantic, things may be different.) We will call him Fitzwilliam (Bill) Darcy.

Liz and Bill are the backbone of the simulation. They remain with it from start to finish. There are several other simulation leaders, who usually work with a particular assistant. The other members of the team are specialists who join in as they are needed, do their work, and move on to the next simulation.

The first specialist is, of course, Doctor Whom. Doctor Whom is probably the only person around who has a postgraduate qualification. He is responsible for the overall running of the simulator, for planning simulation runs, and, from experience, estimating how many training runs may be required, and when to move on to 'measured' exercises. (The quotation marks are here to emphasise that all exercises should be measured, both to ensure that the participants are taking them seriously, and to find any flaws in the measurement plan before it is too late.) He is also present at all simulation runs, and is responsible for the smooth running of each simulation run. He may have to stop a simulation run in mid-flow for technical reasons, which he then must justify to Liz Bennett.

His first assistant is known as Grumpy. He has the task of making the simulator do what is required of it, in terms of hardware and software. Since he is frequently faced by undocumented errors and incomplete specifications, and has to sort these out to a restricted time scale, he usually lives up to his name. One unfortunate holder of this position told me that, by working weekends, cancelling his holidays, and working till midnight six days of seven he had just succeeded in bringing an entirely new version of the simulator software on line one week before a Very Important Simulation. The Director congratulated him in the words: "Since you have got it done early, you must be overstaffed – I'm moving your best programmer to the accounts department."

Dopey is the data collection specialist. He claims his name reflects his giving the straight dope on the results. He designs the experimental plan that he hopes will be used, bearing in mind that changes may be required *en route*. He is responsible for analysing the data collected during the simulation about the actions being simulated, in more or less detail, as required (see chapters 9 – Targets of Measurement and 10 Recommended Measurements). He is responsible for on-line analysis, 'quick look analysis' immediately after runs, and final overall analysis (Chapter 13). He may also have to run statistical data reduction, analysis and presentation in the forms required by the Simulation Leader. (Chapters 13 and 14).

Sleepy is an ergonomist or human factors specialist. He is not always present, (or as a Dutch colleague expressed it, he is not all there.) He devotes himself to trying to make 'objective', convincing and accurate measurement of the strain experienced by controllers. He is responsible for the design of questionnaires, in which participants express their opinions of the simulation. His life is a lot easier since he discovered 'Survey Monkey'.

Grandad heads a group of experienced ex-controllers and assistants. They act as the controllers of sectors adjacent to or under those being simulated. They are known as 'ghost', 'feed' or sometimes 'dummy' controllers. One 'feed' controller may control several adjacent sectors. Their main task is to ensure realistic 'handovers' of traffic entering or leaving the simulated area. They are not really required to provide realistic control, although they take a pride in handing over traffic at the correct time and place, properly separated. Grandad is an atypical feed controller'. He is actually a qualified airline pilot. After a career with a large commercial airline, he was obliged to retire due to ill health, while nearing pensionable age. He enjoys supplementing his pension by acting as a feed controller. It suits him to work part-time, in office hours and his experience of real commercial flying can be extremely valuable at times. He also helps to preserve Bashful's grip on sanity.

Bashful Bashfulsson, an immigrant from another fictitious universe, leads a hunted life. An unkind fate has placed him in charge of the 'simulator pilots'. These are (mostly) young (mostly) ladies. Their main duty is to fill the place of airline pilots, providing dialog with the participant controllers, reading computer generated instructions, and inserting the instructions given by the controllers. When the simulator is not running they have virtually nothing to do, so are assigned to 'general assistance'. This they interpret to include making Bashful's ears turn red, which they invariably do after about five minute's effort. In additional to minimal secretarial duties and intensive gossip, they file data records, transcribe questionnaires, and do anything else that can be found for them to do and which they cannot avoid.

Jack the Lad is the only male 'simulator pilot'. He is young and has recently qualified as a commercial pilot. Essentially, he is waiting for the recession to ease sufficiently for him to be hired as a junior pilot. In the meantime, he is gaining experience that may later be useful, making friends among controllers and learning to understand their ways of thought. He leads a very active social life, and is occasionally, particularly on Monday mornings, in a less than optimal condition.

Ditsy is a more typical 'simulator pilot'. She is experienced in simulation, not only in the ATC context. Sometimes she is inclined to 'explore' the limits of the simulator. On one memorable occasion, the simulator appeared to come to an unplanned halt, but in one sector only. The remaining sectors appeared to be functioning normally. Eventually, someone noticed that Ditsy had systematically reduced the cruising speed of all aircraft in the sector to zero, "to see what would happen". She found out. (She had been planning to leave, anyway.)

Wendy is a rather less colourful type, since she represents the bulk of 'simulator pilots'. She is intelligent, willing, but not technically trained. She has been given practical training in ATC voice communication, and conscientiously follows the standard procedures. One or two 'simulator' pilots provide the voices of all the aircraft in a particular sector.

Although all modern commercial aircraft have two pilots, it is not usual to have two 'simulator pilots' to represent the same aircraft, although changing which pilot speaks to ATC is an occasional source of confusion in real ATC. Usually, the telephone lines used to represent the frequency provide a clear, noise-free channel of communication, and the 'simulator pilots' acknowledge and follow controllers' instructions to the letter. Participant controllers sometimes remark that this is unrealistic. Although it is technically possible to degrade the quality of the communication link, it is rarely done. Equally, it would be technically possible to disguise the 'simulator pilot's' voice so that different aircraft appeared to have different voices, but this also has not been done. It is rather more difficult to script realistic arguments between pilots and controllers. On one occasion, in an attempt to generate strain in a controller, the 'simulator pilot' was instructed to systematically misread and wrongly repeat back orders. Although the supervisor, who had not been warned, nearly had a fit, the controller's heart rate did not vary measurably, nor was there any noticeable variation in his speech patterns. (After about 15 minutes, he raised both hands above his head and shook his clenched fists, without interrupting his calm, clear speech.)

Participants

The most important people in a simulation are the participants. In ATC these are Air Traffic Controllers. Experimenters coming from an experimental psychology background may be tempted to refer to operators as 'subjects'. This should be carefully avoided, since it conveys a very wrong idea of what simulation is about. Operators are not an undifferentiated flock of average citizens. They are experienced individuals, who know a great deal more than the organisers of simulations about the activities being simulated. They often have difficulty in expressing their knowledge, so that they can say that a proposal will not work, but find it difficult to say why.

Normally, controllers working in Air Traffic Control centres form stable teams, who work together over weeks or months and develop an intimate knowledge of each other's' strengths and weaknesses. These teams - sometimes called 'watches' - develop their own individual styles to suit their particular circumstances. Their common spirit can be remarkable. In fact, one female controller in a certain European country took four months off to have an unofficial baby, while apparently never missing a shift. The controllers concerned believe that management was unaware of what was happening.

Ideally, several teams of controllers working with their familiar partners would test a proposed re-organisation. In practice, the pressure on staffing is usually such that seconding an entire watch to participate in a simulation is out of the question. Scratch teams, made up of volunteers from different watches or even different centres are usually assembled. It is not unknown for staff to be recalled in the middle of a simulation, with unfortunate consequences for any balanced design. Even where management can be persuaded to leave the planned staffing alone, there is always the possibility of genuine illness. It has even been suggested that participants should be drawn from controllers suspended from duty for medical or disciplinary reasons – hardly a representative group.

Chapter 6 – Defining a Simulation

The first and most critical part of any simulation is deciding exactly what it is for. All too often, the initial proposal contains a very general phrase, which could mean almost anything. While this may save argument while the proposal is being agreed, it invariably creates problems in planning the actual exercises, and may lead to disappointment with the results.

It is therefore very important to ensure that the agreed simulation definition should contain clearly stated meaningful aims.

A few examples: -

- To validate a modified handover procedure in peak traffic
- To verify that Sector NS will be able to handle the estimated 2018 peak traffic
- To define a recovery procedure in the event of GPS failure
- To estimate the workload required by the adoption of direct routeings in Erehwonian Airspace.

And a few counter-examples: -

- To optimise control methodology
- To assure completely safe operation
- To confirm that no improvement is necessary

Although an exhausted planning group, thinking about catching the last flight home may be tempted to accept these political 'motherhood and apple-pie' statements, it will cost dearly later on. Unfortunately, it is not always the people who make the mistake who pay the cost. (Cynics may say that it never is.)

Ideally, there should be a set of questions that the simulation report should answer. It is a mistake, however, to assume that simply answering, for example, the question: -"Can Sector NS handle the estimated 2018 peak traffic?" with a bald NO! is sufficient. The sponsor will reasonably want to know why, how it failed, and what to do to rectify the situation. The project leader will have to be prepared to provide answers to these questions, and it will not be sufficient to say the questions were not in the simulation definition, or that 'further studies are needed'.

In the early days of computing, some organisations considering the introduction of a computer to replace their parchment and quill pens found that the initial design study showed that they did not need a computer, but a reorganisation of their system would have the same effect. Similar things have happened during simulation definition, where a detailed examination failed to produce any question needing a simulation to be answered. On at least one occasion, it emerged that country XXXXX had not had a simulation for some years, although they were contributing (financially) to EREHWONTROL, and the XXXXX Minister for Aviation had to show that they were getting something for their money. Discretion draws a veil over subsequent events.

A very experienced simulation leader told me that he had never received a simulation definition that did not have at least one major omission. It is therefore necessary for there to be some person who can resolve these omissions, preferably from the sponsor's side. Ideally, this person should be physically located at the simulation centre, but with modern communications technology, telephone, e-mail and SKYPE communications might be sufficient.

To help the more practically oriented reader, and for fun, we will invent and follow through an example of a simulation. Although some readers may think they recognise some characters, this is purely coincidental. (Fiction is constrained by the need to appear realistic, while reality is not.)

Example

Ruritania is a nation state situated in the western part of the continent of Erehwon, bordered on the north by Kennaquhair, on the east by Datong, on the south by Nephelocccygia and on the west by Atlantis. It is the main producer and exporter of coprolite, increasingly in demand for international relations. Further details can be found on Wickipedia. (This is true, which is more than can be said for the rest of this example.)

The main airport is Strelsau International. Ruritania is a member of Erehwontrol, which coordinates the Air traffic Control services of the nations of Western Erehwon. In view of increasing international traffic, and tourism, the airspace of Ruritania is becoming overloaded, and some action must be undertaken to relieve the controllers.

The Ruritanian Ministry of Aviation is considering splitting Ruritanian airspace into two sectors instead of the one existing. They are concerned that the additional co-ordination workload may balance out the reduction in traffic under control.

In addition, they are interested in evaluating a computer-based sequencing aid for departures and arrivals at Strelsau International, avoiding conflicts with local traffic from Hentzau and Zenda airports. The procedure, developed by the well-known Artificial Intelligence expert, Dr V. Frankenstein, needs to be validated.

At a meeting of the Erehwontrol steering committee it was agreed that a Real-Time simulation should be mounted at the Erehwontrol Experimental Centre (EEC) to evaluate the proposed airspace re-organisation and the introduction of the Digital Universal Monitoring and Mediation Kontextual Operational Planning Function (DUMMKOPF).

A meeting was accordingly convened at the EEC, headed by the Director, M. M-J Garrotte. Present were; Elizabeth Bennett (EEC), Fitzwilliam Darcy (EEC), Rupert Hentzau (Ruritanian Air traffic Control), Igor Igorovitch (Assistant to Dr V. Frankenstein, University of Strelsau).

It was agreed that a simulation would take place at the EEC, to be designated Ruritania I.

Ms. E Bennett would be Simulation Leader, and Mr. F Darcy Assistant Simulation Leader.

The primary aim of the simulation would be:

- *To determine whether the proposed split of Sector RU into two sectors RE and RW would provide a significant increase in capacity.*

- *To determine whether the proposed split would be effective for peak weekday and peak weekend traffic, as foreseen for 2020.*

An additional aim would be:

- *To validate the proposed DUMMKOPF method for sequencing departures and arrivals at Strelsau International.*

It was also agreed that I Igorovitch should be responsible for installing DUMMKOPF on the EEC simulator, and that R Hentzau should be responsible for assembling suitable traffic samples, liaison with the Simulation team, selection of participants and any other Ruritanian aspects of the simulation.

Chapter 7 – Data

A Real-Time Control room simulation can be considered as an extreme data reduction exercise. Megabytes of data are generated in the form of data files describing the simulated environment, and the dynamic traffic. Controllers spend hundreds of work-hours reacting to this data, and further megabytes of data record the reactions and their consequences. The data are reduced and systematised, analysed statistically, collated with participants' comments and project leader's observations. Ultimately, a report is produced, with great detail surrounding a single bit of information – "Yes, it works" or "No, it doesn't". In reality, the answer is usually rather more complex, including the reasons why it does or doesn't work, suggestions for solving the problems found, and reports of possible alternative methods.

Although, in some sense, practically all of this book concerns data, we can break this down into different types of data playing different parts in the process.

A digital simulator can be seen as a set of layers of data and data handling processes.

Static 'World' Data

An ATC simulator will contain basic data on, for example, the names and positions of all the beacons in the area simulated, the predesigned airways, the performance to be expected, or simulated, of aircraft types, among other things. The level of detail involved will vary according to the context.

As a start, the familiar shape of Ruritania, together with the existing Airspace allocation is available. As is well known , Zenda, Strelsau and Hentzau are roughly in a North-South line, underlying the airway Upper Rose Seven.

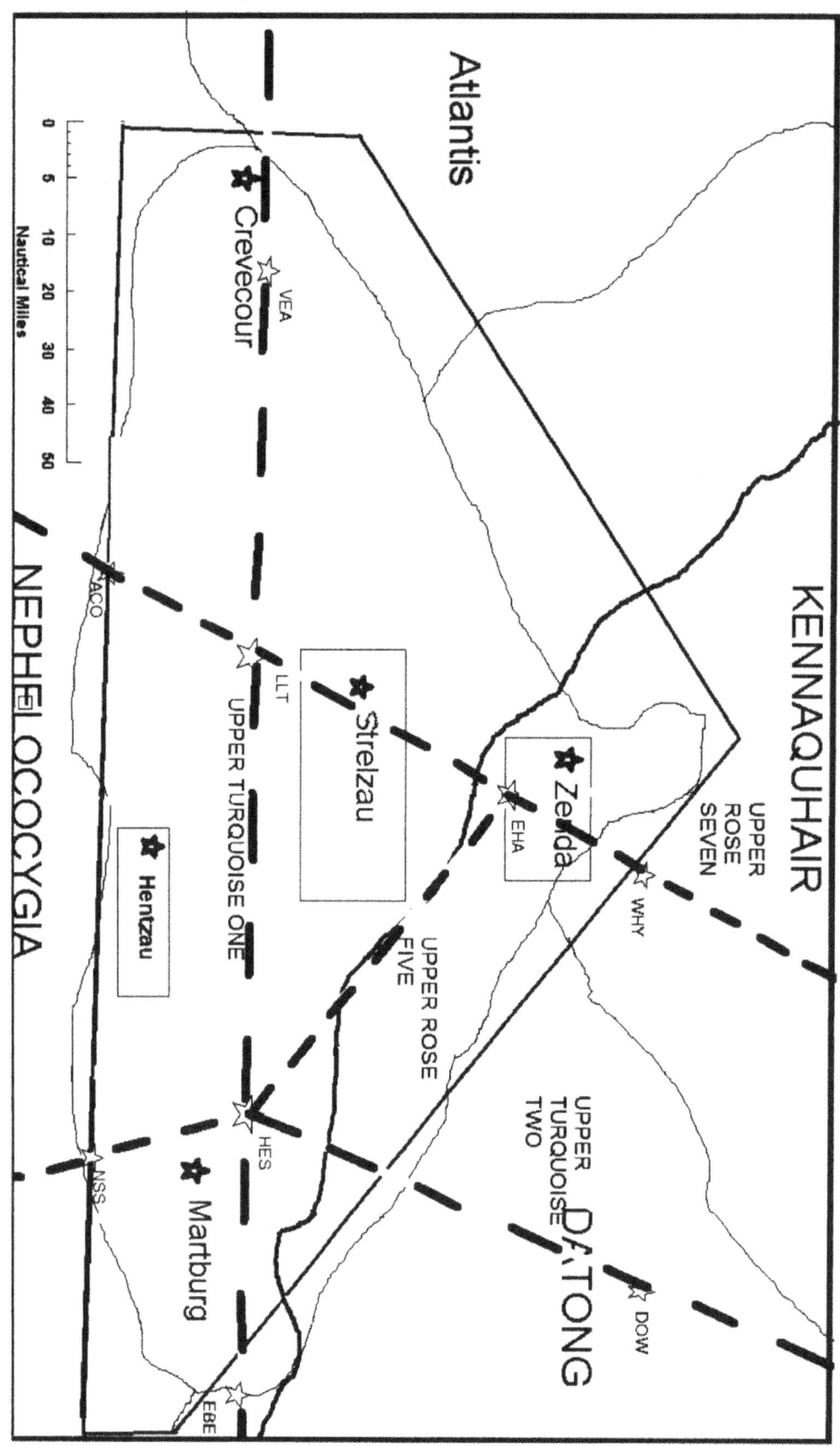

Figure 1 Ruritania Airspace

The initial proposal for two sectors is a simple West-East split.

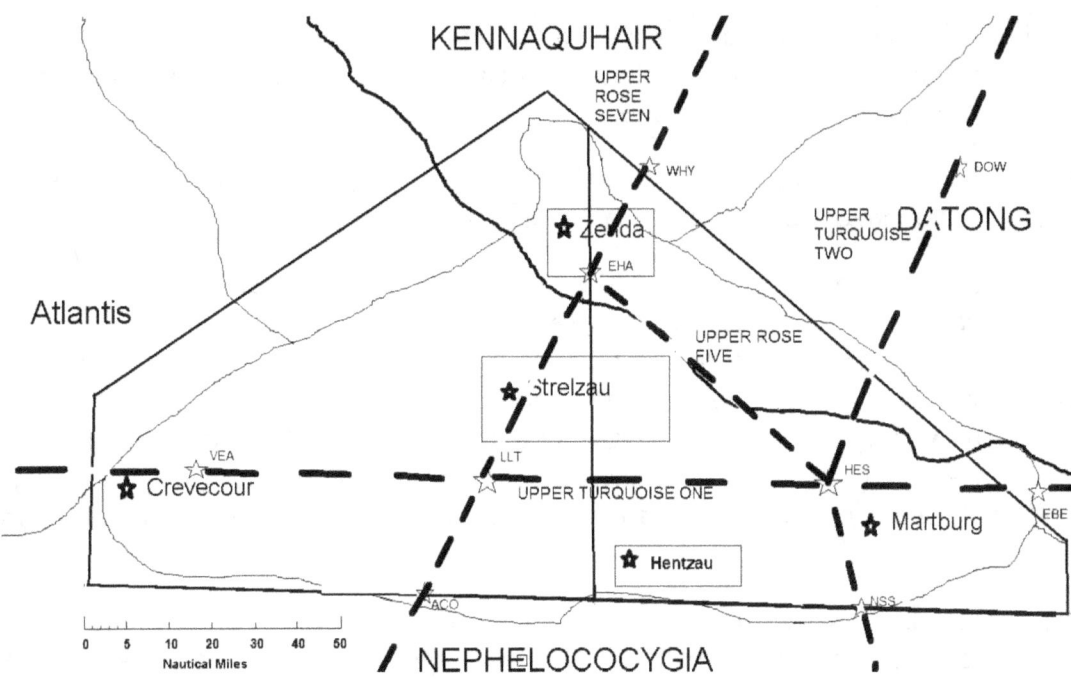

FIGURE 2 Two Sector Split (1)

A brief study of the traffic shows that most UR7 traffic transfers to UR5, as does most UT2 traffic. A second proposal for two sectors gives a more balanced split for traffic, although the ground area of the two sectors is less balanced.

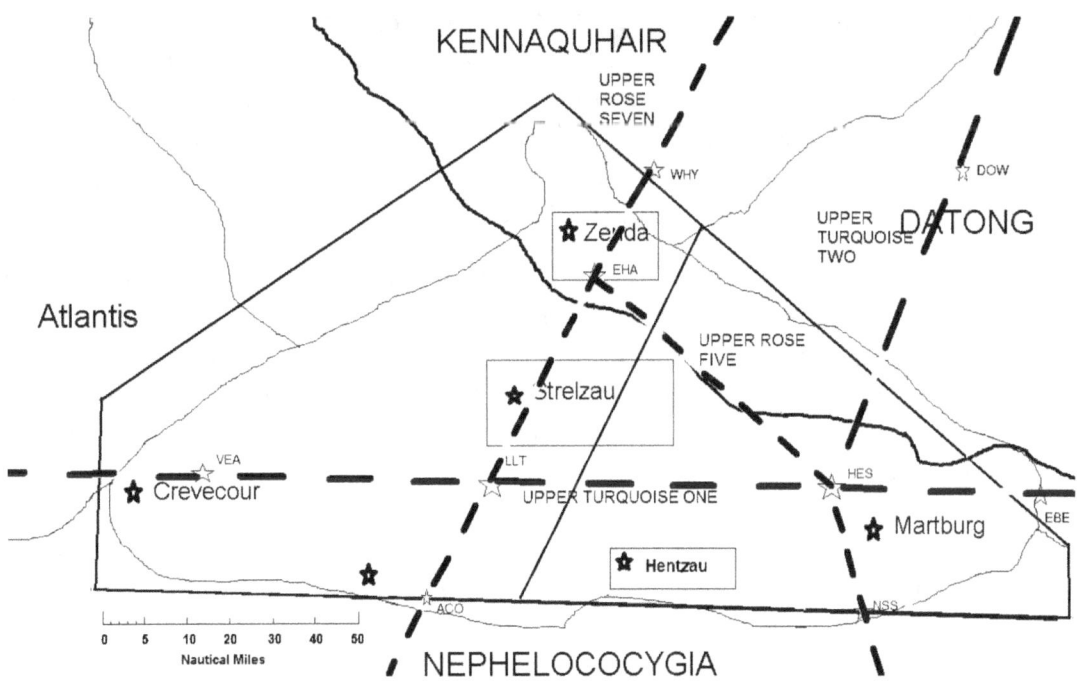

FIGURE 3 Two Sector Split (2)

Dynamic 'Traffic' Data

For each run, a sample of traffic is prepared. (In Air Traffic Control, these samples are essentially lists of aircraft call-signs, with an associated type, a route (defined by beacons and airways), a planned start time (either departure from an airport or entry from an adjacent area.)

Traffic samples are a frequent cause of disagreement between controllers and simulation sponsors, with the project leader caught in the middle.

It might appear reasonable to take the traffic recorded in on a particular day as a representative sample. However, it quickly becomes apparent that there is no such thing as a typical day. Something unexpected is always happening somewhere. On one occasion, we chose an apparently normal day, only to discover after considerable work had been done, that it was the day before the signing of the SALT 1 treaty. This was the first nuclear disarmament treaty and half the heads of government of the developed world were converging on a normally quiet Finnish airport, requiring special clearances and extraordinary flight restrictions on everybody else.

There is an apparent conflict between 'real' and 'realistic' traffic. In practice, this conflict can usually be resolved with a little forethought. There may be typical patterns of traffic associated with, for example, week-ends or public holidays, contrasted to a 'normal weekday'. In fact, a truly representative sample is rarely very useful, since the system must be capable of handling 'reasonably exceptional' loads, rather than average loads.

Traffic samples may also have to be chosen to test alternative 'organisations'. If the traffic is too light, both organisations will cope easily. If it is too heavy, neither organisation will cope. The 'difficulty' of samples containing the same mix of traffic is hard to measure objectively, particularly before the simulation. If the same sample is used with two different organisations, the controllers will be potentially familiar with it on the second run. They will not necessarily recognise it consciously, but electro-encephalographic (EEG) measures of brain activity show that the process of familiarisation occurs even when controllers do not know it is happening. (David, Farbos, Bourgeois, Cabon, & Mollard, 1999; David, Mollard, Cabon, & Farbos, 2000)

Finally, controllers often express dissatisfaction with samples without being able to indicate exactly what is wrong with them. In part this may be because they do not have the vocabulary to communicate their worries. However, a social anthropologist who studied several simulations suggested that controllers feel they 'own the traffic', so that they can complain about it. (Dubey, 2000). The actual source of their unease may be something completely different. They may, for example, think that a procedure will be dangerous, but not want to appear timid or workshy. They may fear that a change may lead to understaffing and consequent overloading. Conversely, they may fear that a change may lead to overstaffing and consequent redundancies. Strangely, controllers do not always feel that supervisors can be trusted to solve such problems.

A series of 'traffic samples' is usually constructed, in the hope that differences will be ironed out by the 'law of averages'. This can involve an enormous amount of painstaking work, requiring familiarity with the airspace and the normal traffic. An attempt at 'scientific traffic generation' was a fiasco. Among other curiosities, it generated a flight of three USAF heavy freighters attempting to land on a grass strip employed by helicopters in the Vosges mountains. The system was literally laughed out of court.

A pragmatic alternative is to take a sample of real traffic and double it by inserting duplicate flights delayed by 5 to 15 minutes, with altered call signs. Samples can then be generated by deleting random aircraft to produce samples with 50% to 150% of the original number of flights. The departure times should be subject to a random perturbation between +5 and -15 minutes corresponding to the uncertainty in real life. An adequate number of sufficiently different samples can be generated in this way. Since day to day traffic is repetitive in any case, this is more realistic than purely random samples would be. A specialised program should be available to check for flights initially in conflict. The feed controllers will resolve most of these conflicts, but they can be overloaded.

Rupert Hentzau returned to his office in the Ministry building at Strelsau, and began a flurry of phone calls and e-mails.

He called up recent records of traffic in Ruritanian airspace. He set an assistant to list the aircraft types currently in use. Before leaving the EEC he had gone to the data section where an obliging data analyst , Wentworth, gave him a copy of the general data book issued to controllers, listing all the types of aircraft for which data was stored in the EEC simulator. A week or so later, the unlucky assistant reported that 134 of the 138 types she had found were in the EEC database. Two of the others turned out to be variants of existing types under different commercial names, and two were vintage aircraft, which Hentzau decided to omit.

Another assistant transferred the computer files of traffic to an EXCEL database and derived a set of data representing the 95%ile levels of traffic on weekdays and weekends in Ruritanian Airspace. From this database, Hentzau extracted some summary tables showing the traffic loads at the three major airports, Hentzau, Streltzau and Zenda on an hour-by-hour basis. He prudently stored these, for use in the final report.

Being an essentially visual thinker, he plotted these hourly rates – which appear as figures 2 to 9 in the report. Figures 2-5 showed that the peak hours for weekly traffic occurred from 0800-1000 and 1600-1800 for all three major airports in Ruritania.

Figures 6 to 9 showed that the weekend traffic pattern was very different. Strelsau International was relatively quiet, but both Hentzau and Zenda experienced peaks of arrivals from 0800 – 1000 and of departures from 1600 – 1800.

Hentzau then called up several samples of weekday and weekend traffic over the relevant periods. He sent these off, with a brief cover note, to the EEC and went home for a well-earned weekend of dissipation.

Chapter 8 – Experimental Design

All trained experimental psychologists, ergonomists, and human factors engineers learn about experimental design early in their training, and most will have at least one book on the subject on their shelves. As undergraduates, they will have endured or enjoyed courses and practical experience with the design of experiments. Other engineers, technologists and system designers will have more or less familiarity with the techniques employed in the less exact sciences to extract the meaning from data containing much unavoidable 'noise'. Rutherford's dictum "If your experiment needs statistics, you should have designed a better experiment" simply does not work when human beings are involved.

The ancient Greek Heraclitus said, "You cannot bathe twice in the same river", or words to that effect. The second time you do something, you know at a very basic level what to expect – compare driving to an unfamiliar destination with driving there a second time, or the first and second times you made love. Human beings are not equipped with a RESET button, so that the stress of a new experience can never be reproduced exactly. If you are trying to compare, for example, two different ways of presenting data on a screen, using monochrome or coloured displays, you might prepare a series of questions about the data and time the operator's responses, checking for errors as you do so. If you use the same set of questions for the two displays, you will usually find that the second set of responses will be faster and more accurate. The experimental psychologist may deal with this by measuring the responses of hundreds of operators, giving the operators practice on similar sets of questions and varying the order of presentation of the displays. She may also try to match pairs of operators who share as many characteristics as possible, or split the set of questions into two sets of similar questions and permute these for individuals or groups of subjects. She can then use classical statistical analyses to separate the effects of type of display, order of presentation and sets of questions, and define a level of unexplained variation which can be called a 'residual' with which the separated effects may be compared to show that they represent a real difference.

Unfortunately, even in the largest Real-Time Control room simulation, we can call on only a few teams of operators. We are constrained to test only a few alternatives. One organisation is often intended to represent the current situation, with which the observers will be familiar. This organisation has often been used for training and familiarisation before proceeding to measured exercises. It is tempting, but wrong, to begin the measured exercises with the current organisation, and follow it with the alternatives. Since there is never enough time for the operators to learn the simulated system completely, even when the experimenters are trying to copy the current system, there will still be errors, slow reactions and difficulties being ironed out which will produce a bias against the first organisation tried. A balanced order of organisations is preferable for objective assessment of simulations.

However, it is usually impossible to realise the neatly balanced designs provided in statistical textbooks. There are many reasons for this. For instance, the sponsors may want to rotate staff through different working positions to get as many different opinions as possible. Staff may fall ill, or be recalled for other reasons. It may become clear very early that one or more of the alternatives is simply unworkable, so that operators see no reason to continue with simulations merely to provide a balanced design. Sometimes, a new organisation combining the best features of several original candidates can be worked out. If this happens, the new organisation will necessarily be tried out during the latter part of the simulation, when the operators are familiar with the equipment and the traffic. Since they are usually the originators of this organisation, the operators tend to be predisposed in its favour.

Even where a determined analyst succeeds in imposing a properly balanced experimental design, his efforts may be wrecked by system failures. Since Real-Time Control room simulation software is designed by a handful of systems analysts and programmers, in contrast to the hundreds involved in producing the real system, the software is put together without painstaking safety checks and procedures are rarely tested to any significant extent. Simulation runs are therefore liable to stop abruptly at unexpected times. The project leader is then faced with deciding if sufficient information has been gathered from the run so far. If not, it may be possible to freeze the situation and re-start it (usually based on technical considerations beyond his competence). If the freeze lasts more than fifteen minutes, it is usually necessary to abandon the simulation run. Sometimes, if the exercise is in its early stages, it can be re-run from the start. Otherwise, the exercise must be abandoned and an alternative version must be run with a different traffic sample.

The hapless analyst (Dopey) will find himself developing experimental designs on the spot, trying to balance samples, organisations, and their order of presentation. In the face of all these difficulties, it is tempting to give up the attempt to design a simulation at all. The actual participants will probably not care, but it is all too easy to end up with a situation where it is not possible to sort out the effects of different organisations from the order in which the organisations were presented. (This is technically referred to as a confounded nuisance.) A useful practical rule is to ensure that the first traffic samples and organisations are re-run at the end of the simulation. This provides at least an indication of whether learning is still taking place. Altering the call signs will usually ensure that the participants do not recognise the samples.

Dopey entered his office at EEC at nine thirty on Monday morning – this is, after all an international organisation – and by ten was ready to start work. On opening his e-mail he found the usual offers of penis extenders and prizes in lotteries he had never entered, but also a note from Rupert Hentzau and a stack of files in Dropbox.

Hentzau's notes showed that there were traffic peaks about 0900 and 1600 during the week, affecting all three airports, although Strelsau International was most heavily loaded. At weekends, Hentzau and Zenda both had arrival peaks in the morning at about 0800 and departure peaks in the evening around 1800.

Ideally, this suggested that there were four peak times to be simulated, each with two potential organisations – with one sector as now, or with two sectors. Since the major question was what capacity could be reasonably expected, a range of traffic samples would be needed. The four peak times could be called WkAM, WkPM, WeAM and WePM. The one or two sector organizations can be marked with 1 or 2 in front of the peak time, giving   1WkAM, 2WkAM,1WkPM 2WkPM, 1WeAM, 2WeAM, 1WePM, 2WePM eight different situations in all.

These would need to be run with different levels of traffic. Starting with 50% level training samples, 80%, 100%, 125% and 150% should, Dopey thought, cover the range of possibilities. Five levels of eight cases made up 40 runs, minimum. At four runs per day, five days per week, that would amount to two weeks. From bitter experience, Dopey knew this meant a good three weeks.

At 1200 hours Dopey went for lunch, collecting Sleepy, Grumpy and Grandad on the way. Over lunch, they talked about Ruritania, from the ATC point of view. Grandad reckoned that they really needed about 150 minutes to simulate 120 minutes of traffic, allowing for a build-up from an empty airspace. Dopey and Grumpy agreed that four 150-minute runs, making up ten hours of working time, with no allowance for setting up or de-briefing, was simply too much. Dopey asked if they could cut it down to 120-minute runs, which would still require eight hours per day or even 90-minutes runs, bringing the running time down to six hours. Grandad asked whether there would be one or two teams of Ruritanian controllers, and how about another bottle. Motion carried unanimously, with the rider from Sleepy that someone would wake him up if - and before - anyone important turned up that afternoon.

Back at the office, Dopey found a text from Liz Bennet, saying she and Bill Darcy would drop in to discuss the experimental plan 'this afternoon'. Since it was already 1500 hours, he made a rough plan, assuming four 90-minute runs per day. He allowed two days for learning, and two measured days in week one, four measured days in week two and four measured days in week three, leaving three days for emergencies and slippage.

He hurriedly made up an exercise table and printed three copies.

Day of Week/ Session	Monday 12/10 Training	Tuesday 13/10 Training	Wednesday 14/10 Measured	Thursday 15/10 Measured	Friday 16/10 Stand-by
0900-1100	Arrival	2WkAM-80%	1WeAM-100%	2WkPM-100%	?
1100-1300	Introduction	2WkPM-80%	1WkAM-100%	2WePM-100%	?
1400-1600	1WkAM-50%	2WeAM-80%	1WePM-100%	2WkAM-100%	?
1600-1800	1WkPM-50%	2WePM-80%	1WkPM-100%	2WePM-100%	?

Day of Week/ Session	Monday 19/10 Measured	Tuesday 20/10 Measured	Wednesday 21/10 Measured	Thursday 22/10 Measured	Friday 23/10 Stand-by
0900-1100	1WeAM-125%	2WkPM-125%	1WeAM-150%	2WkPM-150%	?
1100-1300	1WkAM-125%	2WePM-125%	1WkAM-150%	2WePM-150%	?
1400-1600	1WePM-125%	2WkAM-125%	1WePM-150%	2WkAM-150%	?
1600-1800	1WkPM-125%	2WePM-125%	1WkPM-150%	2WePM-150%	?

Day of Week/ Session	Monday 26/10 Measured	Tuesday 27/10 Measured	Wednesday 28/10 Measured	Thursday 29/10 Measured	Friday 30/10 Stand-by
0900-1100	1WeAM-125%	2WkPM-125%	1WeAM-150%	2WkPM-150%	?
1100-1300	1WkAM-125%	2WePM-125%	1WkAM-150%	2WePM-150%	?
1400-1600	1WePM-125%	2WkAM-125%	1WePM-150%	2WkAM-150%	Farewell
1600-1800	1WkPM-125%	2WePM-125%	1WkPM-150%	2WePM-150%	Departure

FIGURE 4 - Initial Experimental Plan (1)

When Liz and Bill arrived, they were suitably impressed. Liz was worried that four exercises per day would be too tiring for the controllers. Dopey, thinking fast, said he assumed they would have two teams of controllers. Since the planned date was in October, there should not be much pressure on the real ATC service. Liz was doubtful, but said she would see what Hentzau thought about that.

Looking at the plan, Bill said that Week 3 seemed to duplicate Week 2 - could they do without it? Dopey had been hoping they wouldn't notice that. He explained that the controllers would need some time to 'stabilise their performance'. Liz did not make a comment, buy made a mental note to ask Sleepy about this – not that Sleepy really could give a clear answer to the question – or any other in her opinion. Officer training and academic caution do not always fit well together.

Bill had his own ideas. He thought that it might be well to try a 175% sample on the third Monday and Tuesday. He thought a final 100% sample on Wednesday and Thursday would be a good idea. The controllers should find it a doddle by then, and it would do their morale good. The 150% and 175% samples would have left them a bit limp, he thought.

"DUMMKOPF!" exclaimed Liz. Dopey looked briefly offended, then groaned.

"I'd completely forgotten about that. I suppose we have to try it on Strelsau traffic. Will it affect the traffic?"

"Not if it works – but will it?"

"If we try it first on Strelsau weekends, they won't be too loaded at the start."

We'll have to ask about a bit, Dammit,"

Liz summed up.

"Bill, check with Rupert if we can have two teams of controllers. Dopey, will you do an experimental plan with 175% and 100% in the last week, and mark off the DUMMKOPF runs. I'll see what Grandad thinks about the possible effect of DUMMKOPF on the general traffic."

Next morning Dopey took a look at the experimental plan, changing the third week as suggested, adding an "X" for the proposed DUMMKOPF runs.

Day of Week/ Session	Monday 12/10 Training	Tuesday 13/10 Training	Wednesday 14/10 Measured	Thursday 15/10 Measured	Friday 16/10 Stand-by
0900-1100	Arrival	A2WkAM-80%	B1WkAM-X 100%	A2WkPM-X 100%	?
1100-1300	Introduction	B2WkPM-80%	A1WkAM-X 100%	B2WkPM-X 100%	?
1400-1600	A1WkAM-X 50%	A2WeAM-80%	B1WePM-X 100%	A2WeAM-X 100%	?
1600-1800	B1WkPM-X 50%	B2WePM-80%	A1WePM-X 100%	B2WeAM-X 100%	?

Day of Week/ Session	Monday 19/10 Measured	Tuesday 20/10 Measured	Wednesday 21/10 Measured	Thursday 22/10 Measured	Friday 23/10 Stand-by
0900-1100	B1WkAM-125%	A2WkPM-X 125%	B1WkAM-150%	A2WkPM-X 150%	?
1100-1300	A1WkAM-125%	B2WkPM-X 125%	A1WkAM-150%	B2WkPM-X 150%	?
1400-1600	B1WePM-125%	A2WeAM-X 125%	B1WePM-150%	A2WeAM-X 150%	?
1600-1800	A1WePM-125%	B2WeAM-X 125%	A1WePM-150%	B2WeAM-X 150%	?

Day of Week/ Session	Monday 26/10 Measured	Tuesday 27/10 Measured	Wednesday 28/10 Measured	Thursday 29/10 Measured	Friday 30/10 Stand-by
0900-1100	B1WkAM-175%	A2WkPM-X 175%	B1WkAM-100%	A2WkPM-X 100%	?
1100-1300	A1WkAM-175%	B2WkPM-X 175%	A1WkAM-100%	B2WkPM-X 100%	?
1400-1600	B1WePM-175%	A2WeAM-X 175%	B1WePM-100%	A2WeAM-X 100%	Farewell
1600-1800	A1WePM-175%	B2WkPM-X 175%	A1WePM-100%	B2WeAM-X 100%	Departure

FIGURE 5 - Initial Experimental Plan (2)

He added an initial A or B for the two teams. He thought that alternate patterns of ABAB and BABA would make sure no team had two exercises in succession, and each team would have one Morning and one Afternoon session (except the initial training session). At this point he stopped and swore – A was getting all the Weekend and B all the weekday traffic. (And what had happened to the third Thursday? – Lucky that no one noticed.) If weekday traffic was tested each morning and weekend traffic tested each afternoon, the presentation order and Weekday/Weekend would be confounded, but that would not really matter. The exact relative probability for this difference did not matter to anyone – he hoped.

He stored the plan and e-mailed copies to Liz and Bill. He added Dr Whom, Grumpy and Sleepy for good measure.

Next week, he thought, I'll take the Weekday and Weekend traffic and double it, putting in 15 minute delays on one set and introducing random entry time differences. I'll get Grandad to run them through the checker to sort out any initial conflicts. We can use the selector to generate percentage samples. We will never repeat the same sample – if we want a 125% sample twice, we will select a different 125% for the second run. The controllers will recognise some of the traffic, but they do so in any case in the real world.

Time to go home now.

Chapter 9 – Targets of Measurement

Capacity

Measures of capacity attempt to measure the amount of traffic that a system can handle. Although this might appear a reasonable approach in the abstract, it is fraught with problems in reality. One might suppose that it would be possible to start with a relatively low load and increase the load until the system broke down. Although this approach may work with purely mechanical systems, it does not work with systems involving human beings.

There are several reasons why it does not work. The first is that most complex systems do not have a simple input-load-output relationship. Real-Time Control room simulations usually start with empty systems, and build up traffic as the simulation progresses. Controllers are not primarily concerned with the traffic they are actually handling, but with the traffic they can see approaching them. They do not merely react to individual events, such as the announcement by an aircraft that it is entering their airspace. Rather they look at the current situation, and decide well ahead of time what they will do with the traffic as it develops. As they are overwhelmed by the traffic, they simplify their work, adopting safe but uneconomic procedures to save thinking time. When the system finally collapses, the controller will have been running in an unnatural operating mode, sometimes described as having 'lost the picture' for an indefinite time. (Occasionally controllers 'lose the picture' but recover before the system collapses, so attempting to identify exactly when the controller loses the picture is not only very difficult, but pointless.) Anecdotally, many Air Traffic Controllers are convinced that major mistakes often occur just after peaks of traffic load, as the controllers are 'unwinding' from their peak efforts. While this description is derived from Air Traffic Control, it applies equally to many other systems, mutatis mutandis.

A second reason is that in most complex systems, the actual workload is not a simple function of the traffic involved. Some samples of traffic may be very easy to work, while others may involve particularly tricky situations. A skilled operator sees problems coming, and will often solve several potential problems with a single well timed intervention..

If traffic samples are devised with increasing traffic from start to end, the actual point of system collapse is extremely variable, so that it is difficult to determine what the maximum traffic rate is without using at least half a dozen different samples, which costs far too much in time and effort.

If, however, steady-state traffic samples are used, some will lead to collapse and others, with the same load, will not. As the operators become more familiar with the system, they adapt their ways of working, so that a 80% traffic load may lead to collapse, while a 125% load a few days later does not.

A further practical point is that collapse usually occurs at one working position, so that the overall information gathered tells you little about the other operators' state. Further, since operators are inherently team workers, they will 'carry' an overloaded colleague, by making extra effort to sort out potential problems before they reach him, or accepting less than ideal output 'downstream' from the sufferer. They do this normally in working control centres to cover for hangovers or 'small baby syndrome'. (This is the chronic fatigue experienced by parents of small babies, particularly evident in countries where maternity leave is not legally provided - currently Cote d'Ivoire, Oman and the United States of America.)

An experienced simulation leader can tell when the traffic level is approaching the breaking point, and will intervene to prevent the more distressing stressing of the participants.

A slightly different technique has been used in some simulations and even in real-life experiments. Here the operator is asked to perform an additional 'secondary' task as well as his main one. Typical tasks are simple addition sums, binary choices or recognising an unusual note in a sequence. The idea is to measure 'spare mental capacity', and is derived from elementary experimental psychology studies. Unfortunately, this simple model of human activity simply does not work when applied to complex tasks involving forward planning, recognition and experience. It appears to have fallen into well-deserved neglect, although it is periodically resurrected as the unwritten lessons of the past are forgotten.

Sometimes it is possible to identify 'embedded secondary tasks' which are legitimate tasks, genuinely part of the job, which can be delayed or even neglected under pressure. In Air Traffic control, for example, verbal handovers should take place about five minutes before the aircraft leaves the sector, but modern systems have automatic hand-over routines that are triggered at the sector boundary. When under time pressure, the controller may neglect routine handovers, knowing that the system can cope with these. This behaviour can be identified by observation or analysis of records, and can form a useful 'objective' indicator of strain. Finding such tasks requires a detailed knowledge of the system, There is no guarantee that they do exist in any case.

It is therefore reluctantly necessary to try running with increasing traffic samples, but to be prepared to halt the simulation run if the system is going into overload, and reassure the controllers involved as soon as possible to reduce the psychological stress they will be feeling.

System Performance

Measures of system performance measure the effort required to operate the system. There are two main types of system performance measure, the objective and the subjective.

Objective measures look at, for example, the number of orders required per aircraft, or the time taken to communicate in seconds per aircraft. The list of potential measures in air traffic is enormous – see below.

Subjective measures generally rely on assessments by the participants, or occasionally by experienced observers, of the difficulty they experience. This apparently straightforward approach has some justification – if controllers say a system is unworkable, they will not work it. However, it is not always convincing to sponsors committed to a solution. Responses may be distorted by the participants' desire to please the experimenters - or to annoy them. They may feel that they are expected to make a particular reply. They may not wish to admit they cannot make the system work. They may try to reflect what they feel the group's opinion is, or ought to be, rather than their own.

It is practically impossible to prevent participants discussing the simulation, and answering questions from other interested controllers. This means that replies are rarely truly independent and sometimes not reliable measures. Knowing exactly when to trust replies is a major part of the craft of simulation.

Operator Performance

Measures of operator performance measure the extent to which the operator controller has carried out the task. The simple avoidance of disaster is not usually sufficiently sensitive a measure. It is usually necessary to establish a norm to which operators' performance can be compared. In Air Traffic Control the relative frequency of conflicts is the most obvious measure. However, as a working rule, if there are enough conflicts to produce statistical significance in the length of time available for a simulation, the controllers are working in an unrepresentative crisis mode.

It may, however, be possible to establish a 'standard performance' or norm, which represents the best reasonable performance, and compare operators' actual performances to this norm. Establishing such a norm is a difficult process, requiring a detailed knowledge of the real task.

Any proposed measure of performance should satisfy certain conditions.

- It should be reproducible.

- It should be sufficiently sensitive to distinguish relatively subtle differences in the control activity.

- It should not be intrusive.

- It should be quantitative.

- It should be 'automatable', preferably within the system, real or simulated.

    (The expense and unreliability of manual data collection and the technical difficulties of synchronising external systems make internally generated measures practically necessary.)

For example, David (2004) describes the process of establishing such a norm in Air Traffic Control, largely based on the scoring method used in the ATC game TRACON (Wesson International, 1990), adapted to provide a continuous score rather than a single final value.

A standard performance - assuming no errors or inefficiencies - can be calculated for each sample on a continuous basis. The controllers' performance can be measured on-line. It can be expressed as an absolute figure and as a percentage of the standard performance. (Sometimes controllers can achieve a value exceeding 100%)

There is necessarily some 'lag' in detecting falling-off of performance.

A similar scoring method can be used in other systems, maintaining the general principle of comparing to a standard, although considerable care will be needed, particularly with supervisory systems where there is no routine work. In some industries, similar systems are used in real life to assess performance. The administration, and ethics, of such systems are beyond the scope of this book (Thank God).

Liz did not really need to think about the target of measurement, since the main questions were explicitly in terms of capacity.

DUMMKOPF was a slightly more knotty problem. The first question was did it work? Second - If it did, did it help?

It would soon be clear if it didn't work at all, but if it did, would there be a measurable improvement in Strelsau TMA? Missed approaches or near-misses would be an objective measure, but would there be enough to show a noticeable difference?

"We'll just have to suck it and see" thought Liz. "In any case, we will want the controllers' opinions – if they won't accept it, no statistics will change their minds".

Now I'd better work out what measures we need. ----

Chapter 10 – Recommended Measurements

To impose some order, it is simplest to begin with the most immediate 'on-line' measures, then measures of single runs (quick-look), then overall comparison measures.

On-Line Measures

'On-line' measures are taken during the running of a simulation run. They are usually displayed at the supervisor position, and tell the supervisor how the run is going.

The two most widely used measures are SWAT and ISA.

SWAT, the Subjective Workload Assessment Technique, is a self-reporting technique. The operator is asked to report how stressed (s)he is on three scales – Time Load, Mental Effort Load and Psychological Stress load. Each scale has three levels.

In most circumstances, the operator is asked to give their rating verbally, and the response is noted by a trained observer. Clearly, this is not practical where a dozen or more operators are working simultaneously, if only because there would not be enough space to fit all the observers in to the control room. An automated prompt, with a dedicated keyboard, could be installed if required, but would be expensive.

.Direct experience suggests that SWAT has several drawbacks.

1. The scoring method is complex, and requires a complex computation.
2. The scores are not independent in practice.
3. The rating process interferes with performance, particularly when under stress.

On the basis of practical experience, as an ergonomist and as a statistician, I distrust measurement techniques that rely on mathematical models of human mental performance. They oversimplify the extraordinary complexity of human thought, and are often both psychologically and statistically naive.

A full description is available in Gawron (2000), including more than you will wish to know about its past use, although the scoring system is not described in sufficient detail for reconstruction in the regrettable absence of information from Wright-Patterson AFB.

For a more detailed assessment of SWAT in the context of ATC simulation, see David and Pledger (1995) and the thesis by Pledger (1994). Stanton *et al* (2013) gives a very comprehensive review of the use of SWAT, as does Wilson (ed) (1995).

ISA (Instantaneous Self-Assessment) was originally developed by the U.K. ATCEU (Air Traffic Control Evaluation Unit) at Bournemouth (Hurn) airport. It has been used by other UK organisations, including DERA (Portsdown), the Royal Navy's research organisation - now part of Qinetic. It has also been used by the European EUROCONTROL, the French CENA and by the Dutch RLD.

In spite of the efforts of some would-be scientists, ISA is refreshingly free from theoretical backing.

ISA is usually applied through a computer-based system. Each operator is supplied with a simple keyboard with five keys. At regular intervals, usually about two minutes, but in our example 3 minutes, the keys begin to flash. They continue until the operator pushes one of the keys, or for 20 seconds. The five keys represent the operator's self-assessment of his workload at that time. In EUROCONTROL use, the five levels were marked: -

- VERY HIGH   - Completely occupied, some tasks missed.
- HIGH -   Almost completely occupied, but task can be completed
- FAIR -   Steady, reasonable workload. Some breaks
- LOW -   Little work, much spare time
- VERY LOW - Practically no work, boredom, lack of stimulus

In any general system, I would suggest using colour coding and symbols as below:

- VERY HIGH – Red   + +
- HIGH   - Amber   +
- FAIR   - Green   =
- LOW   - Cyan   -
- VERY LOW – Magenta   - -

The use of a colour code has the advantage that it is language-neutral, and does not give rise to the problems experienced by earnest psychologists trying to define five separate equally balanced levels in not more than a dozen characters each in forty-one languages.

Depending on your circumstances, you may construct a separate hard-wired system, such as "Euro-ISA", a system built into the display software, or even a stand-alone 'app' to be displayed on mobile phones.

In any case, it is necessary to foresee a control system to present the query, usually in the form of flashing lights, record (and time) the response, acknowledge the response by turning off the flashing lights, display the response at the supervisor's console and store the responses (and response times) for later detailed analysis.

Although most psychophysiological measures are either too intrusive or too affected by physical activity, it may occasionally be worth measuring heart-rate using one of the contemporary sports measurement equipment. Even with this type of equipment, the effort required usually means that only one or two participants can be measured.

Quick-Look Measures

These measures are available immediately after the end of a run, in a form that is useful to the simulation leader, her staff and the participants.

Perhaps the most general measure is the NASA Task Load Index.

The **NASA-TLX** is a Task Load Index developed by NASA, and widely used as a measure of the subjective difficulty of a task. It involves six different scales, representing different aspects of the task.

Mental Demand asks how much mental and perceptual activity was required (e.g. thinking, deciding, calculating, remembering, looking, searching, etc.)? Was the task easy or demanding, simple or complex, exacting or forgiving?

Physical Demand asks how much physical activity was required (e.g. pushing, pulling, turning, controlling, activating, etc.)? Was the task easy or demanding, slow or brisk, slack or strenuous, restful or laborious?

Temporal Demand asks how much time pressure the participant felt due to the rate or pace at which the tasks or task elements occurred? Was the pace slow and leisurely or rapid and frantic?

Performance asks the participant to rate how successful they were in accomplishing the goals of the task. It also asks how satisfied the participant was with their performance in accomplishing these goals?

Effort asks how hard did the participant had to work (mentally and physically) to accomplish the level of performance they achieved?

Frustration Level asks how insecure, discouraged, irritated, stressed and annoyed versus secure, content, relaxed and complacent the participant felt during the task?

The participants are asked to mark their ratings of the task on each scale by marking a point on a 'Lickert Scale' with 20 graduations. (A' Lickert Scale' is simply a line with graduations marked along it, and usually a word at each end to remind the participant which way the scale goes. ('Performance' is often marked the wrong way round, so it is advisable to remind the participants from time to time.)

As an index is traditionally a single figure, a scaling procedure involving judging which of each possible pair of scales is more important. This is usually carried out after the task, and produces a weighted mean of the scales. Although the procedure may be applied by a computer program, many users prefer to look at the specific scales, since this may give information more relevant (and understandable) than a single overall index. The authors of the task load index have no objection to this approach. (Personal Communication.)

An example of the NASA-TLX form is provided below in Chapter 11, and significant aspects of the NASA-TLX are discussed in Appendix I.

Exercise Measures

These measures describe the run as a whole, taken from the records stored during the run. Although this list is based on Air Traffic Control, very similar lists could be drawn up for other fields of control activity. In many simulations, extra data will be needed and should be defined as soon as possible. A short name is attached to each measure, for future reference. Although data are usually recorded as individual values, they are usually expressed over short time interval. In Air Traffic Control, a three-minute interval is usually chosen. Because traffic builds up at the start of a simulation run, the first fifteen minutes of traffic is discarded.

The following variates are usually derived for three-minute intervals.

The number of aircraft entering, leaving or present in a sector is called Ac, and specified as entering, leaving and present.. The exact definition of Aircraft present is the mean number of aircraft present during the defined interval.

Number of aircraft entering	EAc
Number of aircraft leaving	LAc
Number of aircraft present	PAc

The 'frequency' is the simulated R/T frequency, usually one per sector.

| Number of Frequency Calls | Nfr |
| Duration of Frequency Calls as % of time | Tfr |

The 'Intercom' is the simulated link between working positions within the simulated area – this reflects the movement of aircraft from one sector to another or a TMA..

| Number of Intercom Calls | Nin |
| Total duration of Intercom Calls as % of time | Tin |

'Handovers' are calls to and from the Feed sectors.

| Number of Handovers | Nho |
| Total time of Handovers as % of time | Tho |

'Orders' are the instructions entered into the computer system

Number of Orders	Nor
Total Time spent entering orders as %age of time	Tor
No of Incomplete or Cancelled orders	NXo
Duration of Incomplete or Cancelled orders	TXo

For each TMA, some measures of their efficiency must be recorded.

Number of aircraft landing	NLa
Number of aircraft departing	NDe
Seconds delay landing	DLa
Seconds delay departing	DDe

As well as being presented to the exercise supervisor, the ISA responses and the time taken to respond are also recorded at three minute intervals.

ISA response at three minute intervals	VIS
Time to respond to ISA query	TIS

Finally we can add Heart Rate (HR) – although this is usually measured for one or two crucial positions.

Heart rate at three minute intervals	HRt

(For each variate measured at three-minute intervals the mean and peak values are also recorded, using the values from 15 to 90 minutes only, since the first fifteen minutes are unrepresentative because the system starts with no aircraft present. These values have the suffix M and P respectively. Thus MEAc represents the mean no of aircraft entering, while PVIS represents the peak three-minute value for the ISA response.)

Some values are recorded for the total measured part of the simulation run. This is usually because they correspond to events so rare that there are too few to provide reliable three-minute values

The number of 'mild' conflicts – Aircraft approaching within 5 NM – and of 'severe' conflicts – Aircraft approaching within 2 NM are recorded for the period of the simulation. For TMAs the number of Missed Approaches is also recorded.

Mild Conflicts	NC5
Severe Conflicts	NC2
Missed Approaches	NMA.

In addition, the NASA-TLX scores and the index itself may be included'

NASA-TLX value	TLX
Mental Demand	TLMe
Physical Demand	TLPh
Temporal Demand	TLTe

Performance	TLPe
Effort	TLEf
Frustration Level	TLFr

While all these data are recorded, those that are measured at three-minute intervals should be recorded in a file for each exercise, clearly marked with exercise code, date and time. An additional file should cumulate the exercise-length measures (TLX, Peak, Total and mean values) for each exercise, with the exercise code, date and time. These files are best stored in Microsoft EXCEL. The advantage of a widely-used format and programs that are more-or-less debugged outweigh the many defects of this package.

Whatever package is used, system and discipline are essential to ensure that the data are actually recorded. A named person must be responsible, and the files must be unequivocally marked.. *(Dopey, in the Ruritania simulation.)*

Choice of Measures

The newcomer to simulation may well quail before this cataract of data, but some help is at hand. A very experienced Head of the Simulation Division at EEC, the legendary Mr. Charles Massie, devised a simple tabulation where the available methods were given on the Y-axis, and the aims of the simulation on the X-axis.

Examples of the Massie Grid are presented as Figures 6 and 7 following. A question mark shows that this method may be relevant to this aim. An X shows that it is not.

The fourth column in Figure 6 which is added for this simulation, shows how to compare values for the two sector with the one sector alternative – using the sum, average or peak values for the two sectors – or not used (see below Chapter 13 - Analysis of Variance). This problem only occurs where the actual two sector working positions must be compared with the single sector - the TMA working positions are not affected.

Since we have two separate questions, the capacity increase and the use of DUMMKOPF, Liz compiles two different Massie Grids. One applies to the whole area, (Figure 6) and the other to the exercises using DUMMKOPF and similar exercises where DUMMKOPF is not used. (Figure 7)

It is always tempting to ask for all possible measures to be tested in all possible ways, and it will be observed that Liz Bennett has nearly fallen into this trap. In fact, it is often a basic fall-back position to measure everything, even if this means throwing 90% of the results unexamined. After all, computer time is now cheap, compared with thinking time. Unfortunately, throwing-away time can be just as long as thinking time.

The second and third columns refer to the weekday and weekend traffic, where the differences between the measures are important, but their actual values are often just as important. After all, if the number of intercom calls is twice as high in the two-sector case, it may be statistically significant, but sufficiently low that it does not matter.

Measure	1 vs 2 Sectors Weekday	1 vs 2 Sectors Weekend	Sum/ Mean?
3 Minute Measures (+ Peak and Mean overall)			
EAc	?	?	Sum
Lac	?	?	Sum
PAc	?	?	Sum
NFr	?	?	Sum
TFr	?	?	Mean
Nin	?	?	Sum
Tin	?	?	Mean
NHo	?	?	Sum
THo	?	?	Man
Nor	?	?	Sum
Tor	?	?	Mean
NXo	?	?	Sum
Txo	?	?	Mean
VIS	?	?	Mean
TIS	?	?	Mean
HRt	?	?	Mean
Overall Measures			
NC5	?	?	Sum
NC2	?	?	Sum
NMA	X	X	Sum
TLX	X	X	X
TLMe	?	?	Mean
TLPh	?	?	Mean
TLTe	?	?	Mean
TLPe	?	?	Mean
TLEf	?	?	Mean
TLFr	?	?	Mean
HR	?	?	N/A

FIGURE 6 - Massie Grid – Overall

Liz has requested 16 variates at 3 minute intervals and 8 overall measures. Since each 3-minute variate supplies an overall mean and peak value, there will be 40 variates overall, for the Planner and Executive for one sector in half and two sectors in the other half of the runs.

Measure	Hugh DUMMKOPF On/Off
3 Minute Measures	
EAc	X
Lac	X
PAc	X
NFr	X
TFr	X
Nin	X
Tin	X
NHo	X
THo	X
Nor	?
Tor	?
NXo	?
Txo	?
VIS	?
TIS	?
HRt	?
Overall Measures	
NC5	X
NC2	X
NMA	?
TLX	X
TLMe	?
TLPh	?
TLTe	?
TLPe	?
TLEf	?
TLFr	?
HR	?

Figure 7 - Massie Grid - TMA

For the three TMAs, with two working positions each, Liz has asked for 7 3-minute variates and 8 overall variates. Including the mean and peak overall measures, this produces a mere 22 variates for analysis. Although we are only really interested in Strelsau, where DUMMKOPF is installed, we can repeat the analyses for Hentzau and Zenda. (the average of all three TMAs is of no interest to anyone.

Back to Liz Bennet's reflections.

Specific Questionnaire needed for Strelsau TMA then.- Sleepy to write it.

Now what about the measures overall?

We have ISA, so we might as well use it. - check Grumpy

Ditto NASA-TLX – but we do not need the actual index – I never really know what it means, and I expect no one else does either.

End-of-run questionnaire? Redundant if ISA works, except that this is where participants can bring up specific unplanned problems without risk of embarrassment. Should be able to tack it on to the same sheet as the NASA-TLX.

De-briefings after each run. Either Bill or me must moderate. It should take place in participant's room, or somewhere else, so that next run can be prepared in the simulation room.

Final Questionnaire. For all participants – more or less the standard one.

Sleepy's Operator questionnaire. It keeps him happy and could show up something,

DUMMKOPF – Final interview with the participants who have used it. We may get some useful quotes. Might as well try heart-rate recording as well. It will probably just measure movement, but it is time to trot it out, before it gets lost, again.

Anonymising : Pack of 25 – name and number. And a discrete EXCEL File, with password.

Chapter 11 – Preparation

Simulation Room

The simulation room layout should, as far as possible resemble that at the real place being simulated. Unfortunately, this may not actually exist. Sometimes, several functional suites may be required from different control centres. Screens may be necessary to avoid visual, verbal and occasionally violent interactions not normally possible.

Equally, the equipment used should resemble that in real use, as far as possible. Sometimes, there may be problems in real life caused by unreliable or obsolete equipment. It is not kind (or politic) to simulate these, although they may be mentioned in reports to help build the case for replacement.

We cannot be dragging consoles in and out all the time so I suggest we use a fixed layout. We can use two sector sets side-by-side for area control, executive next to executive. We close down one when we are working the single sector organisation.

The three TMA suites should go opposite, facing away, with partitions between them.

The Supervisors should go where they can see into the three TMAs – say between the two sectors. No – in the real world, the sector sets will be side-by-side, and will be talking to each other. They will have to go beyond the sectors, raised for visibility.

Bloody DUMMKOPF will need an observer, but he can just sit behind Strelsau Approach or outside him.

Dummy positions. Four of them, one for each direction. Normally we can probably bandbox them down to two, but the high traffic 150 or 170 may need all four. They can go on the other side of the TMA row, behind the partition.

Let's see how that looks-

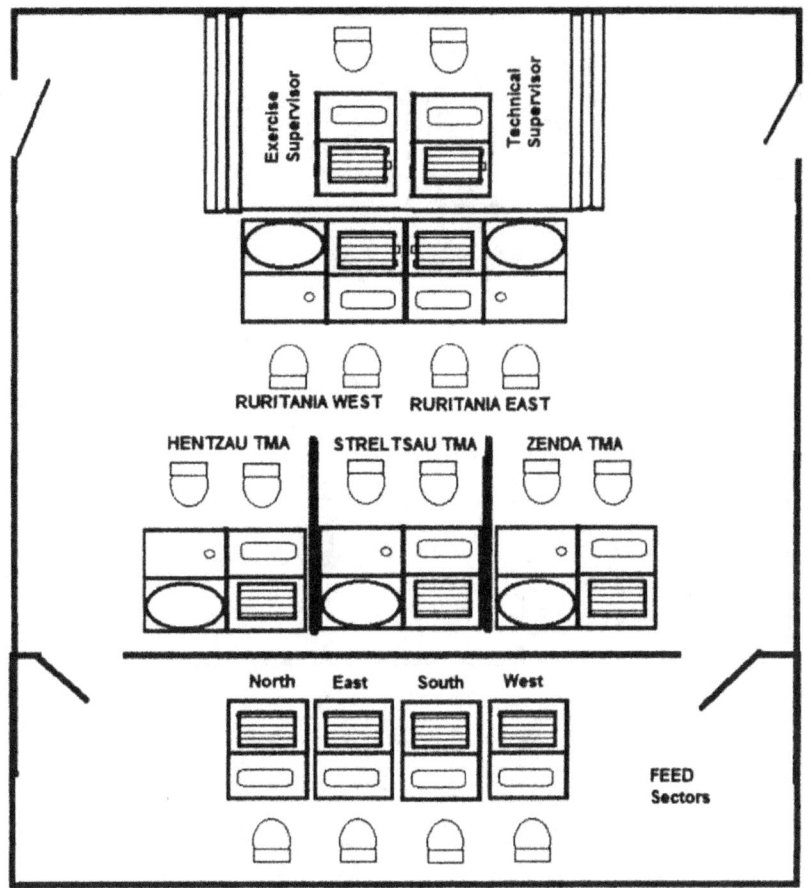

FIGURE 8 - Control Room Schematic(1)

…..Should be OK – I'll send it to Liz and Dr Whom – they can get on with it.

Liz; We need space behind Strelsau TMA for Igor – he'll be sitting in for Dr Frankenstein, I suggest we swap Zenda and Strelsau,

We'll use Ruritania West for the single-sector exercises, so please swap the Exercise and Technical Supervisors.

Dr Whom: Why are we sitting on WCs instead of chairs?

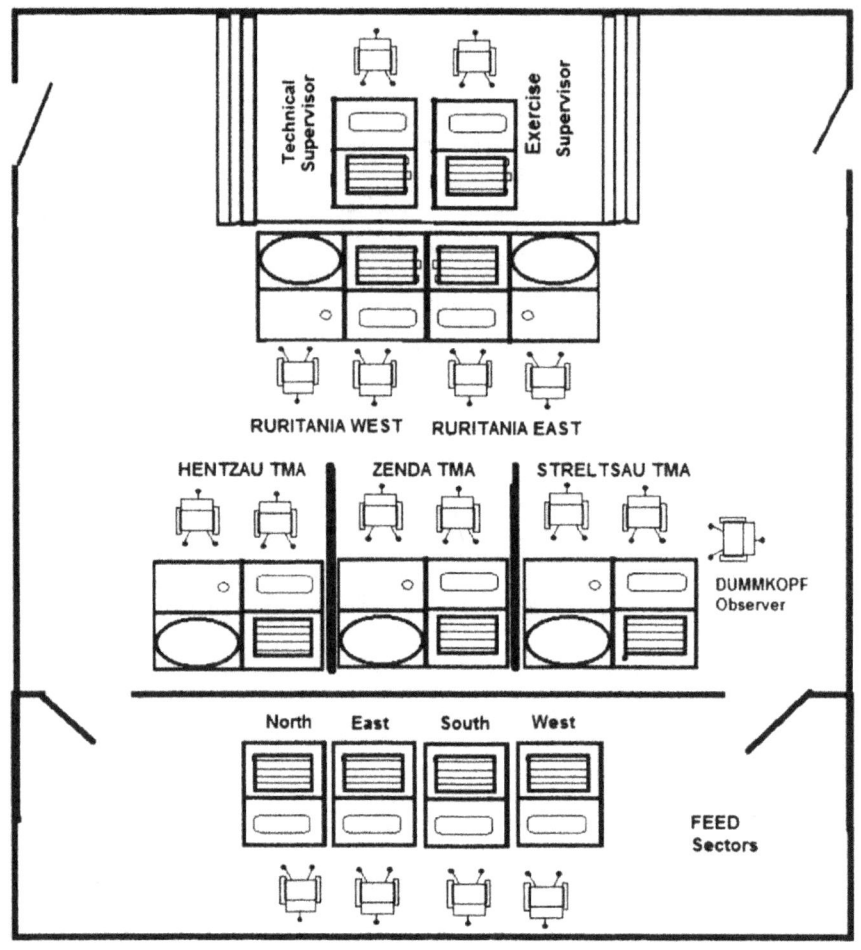

FIGURE 9 - Control Room Schematic (2)

Sleepy : If the last simulation was anything to go by, WCs might be better….

Liz : Ha, bloody ha. Let's leave it at that.

Participant's Room

Most of the preparation has already been described, but the final, very important, part remains. For example, an interview between Bill Darcy and Wendy and Petra, acting assistants out of simulation time.

BILL Wendy, we will need French and German language papers – about half this lot come from the east of Ruritania. Can you order three – no, four – copies each of the Strelsauer Lugenbold and Les Mensonges de Crevecour, for the three weeks of the simulation. Get a lot of ScheissKaffee and six dozen Pferdurinbier and Merdevin.

Halbachtsthal in Utopia City Main Drag are specialists in Ruritanian food and booze – see what they suggest. What sort of biscuits do they like? Find out and get plenty.

Petra, check the room. Get those chairs back from wherever they have wandered off to. And make sure they stay here. Check the cupboards and make sure there are at least three dozen cups and saucers, and that that bloody fridge is working, and this time no bloody mummified mouse in the freezer drawer. How the hell did it get there anyway? That bloody cat gets so much food from you lot that it can hardly walk, let alone chase mice.

Get the cleaners to give the place a good going over. If they do it right give them a tip – I'll pay you back. And can we find some flowers? I think the Deadly Nightshade is the Ruritanian national flower – but make sure there are no berries on it. We will have enough problems without poisoning the waters – literally.

And, Wendy? Check the hotel booking – just in case. And the coach – 12 from the airport and 4 from the main station, at present – the rest are 'own transport' – get them parking permits and, this time, get them back when they leave!

Visitors

One important aspect of a Real-Time simulation is its use to publicise the centre and to show the simulator at work. Visiting parties are usually planned well in advance. Care should be taken to avoid clashing or incompatible parties.

The Director or someone suitably impressive should welcome visitors. They may be shown any available publicity film, and given a brief introduction to the purposes of the simulation – some material can be developed which will save effort later (see Chapter 14 – Reporting). Although most visitors can be expected to speak English, French or German, it is advisable to have an interpreter on hand, if possible. A tour of the more presentable parts of the centre should lead to the visitors' gallery (or area if there is no viewing gallery). It is very important that visitors should not be allowed onto the working floor while the simulation is running. Curiously, the higher the rank of the visitors, the better they react to a firm request not to interfere with the running of a simulation. (These people are serious!) I have spoken to a controller who told someone who got in his way to 'f**** off', only to discover at the end of his shift that he had been addressing Prince Philip. His CO apparently tried to apologise, but was told that the correct form should have been 'f*** off, Sir'.

After the simulation, the visitors should be introduced to suitable participants, although they should not 'sit in' on de-briefings, where they may hear home truths they find unpalatable. More importantly, their presence may inhibit the participants' responses.

Finish with drinks or a meal according to the time of day

The Director phones Liz.

We have a party of visitors from Strelsau

- *Marshal Strakenz and Colonel Sapt from the Ruritanian Army*
- *Captain Von Tarlenheim from the Ruritanian Air Force*
- *Monsieur Ronde de Cuir from the Ruritanian Air Ministry*

They are coming here on the twenty second – Thursday- at 0900 AM, so book them a hotel for the 21st get them picked up in the morning They plan to stay until about 1400. We had better offer them lunch, and I'll want you to talk to them. Work out a program and let me have it. No rush, this afternoon will do.

'Right' says Liz – thinking 'CristiHimmelFart' (Look it up)

'Wendy, reserve four rooms for Wednesday 21 at the Hotel Puteaiselle, or the MetzeAchselHöhleHof. E-mail Strelsau to confirm.'

Liz; So now we need: -

End-of-Exercise Questionnaire
– including NASA-TLX
20 per exercise 56 exercises possible	*1120 in all*	*Print 1200*
End of Simulation Questionnaire	*20 in all*	*Print 25*
DUMMKOPF Questionnaire	*26 in all*	*Print 30*
Operator Characteristics	*20 in all*	*Print 25*

That will keep the printers busy.

NASA-TLX + Observations

DO NOT WRITE YOUR NAME

Relating to this exercise - your personal experience.

Day_____   Exercise Start _____   Participant No.____ Working Position _____

Mental Demand	How much mental and perceptual activity was required (e.g. thinking, deciding, calculating, remembering, looking, searching, etc.)? Was the task easy or demanding, simple or complex, exacting or forgiving?
	LOW |__|__|__|__|__|__|__|__|__|__|__|__|__|__|__|__|__|__|__| HIGH
Physical Demand	How much physical activity was required (e.g. pushing, pulling, turning, controlling, activating, etc.)? Was the task easy or demanding, slow or brisk, slack or strenuous, restful or laborious?
	LOW |__|__|__|__|__|__|__|__|__|__|__|__|__|__|__|__|__|__|__| HIGH
Temporal Demand	How much time pressure did you feel due to the rate or pace at which the tasks or task elements occurred? Was the pace slow and leisurely or rapid and frantic?
	LOW |__|__|__|__|__|__|__|__|__|__|__|__|__|__|__|__|__|__|__| HIGH
Perform-ance	How successful do you think you were in accomplishing the goals of the task set by the experimenter (or yourself)? How satisfied were you with your performance in accomplishing these goals?
Careful!	GOOD |__|__|__|__|__|__|__|__|__|__|__|__|__|__|__|__|__|__|__| POOR
Effort	How hard did you have to work (mentally and physically) to accomplish your level of performance?
	LOW |__|__|__|__|__|__|__|__|__|__|__|__|__|__|__|__|__|__|__| HIGH
Frustration Level	How insecure, discouraged, irritated, stressed and annoyed versus secure, content, relaxed and complacent did you feel during the task?
	LOW |__|__|__|__|__|__|__|__|__|__|__|__|__|__|__|__|__|__|__| HIGH

NOW PLEASE TURN OVER >>>>>

Control Room Simulation

NASA-TLX + Observations (Verso)

Relating to this exercise - your personal experience.

Did you feel at ease with the equipment?   Yes   No

Any comments on equipment?

Was there anything about the exercise you were worried about?

Was there anything else you would like to comment on?

Ruritania I Simulation – Final Questionnaire

Participant No, ____

Please answer this questionnaire from your own viewpoint. Your answers will not be disclosed to anyone outside the simulation. The identity of any respondent quoted in the final report will never be disclosed to anyone outside the analysis team. In particular, this applies to any Ruritanian personnel at any level.

Overall, how satisfied were you with the way the simulation went?

Poor |__|__|__|__|__|__|__|__|__|__|__|__|__|__|__|__|__|__|__|__| Good

Was it generally realistic?

No |__|__|__|__|__|__|__|__|__|__|__|__|__|__|__|__|__|__|__|__| Yes

Was the traffic realistic?

No |__|__|__|__|__|__|__|__|__|__|__|__|__|__|__|__|__|__|__|__| Yes

Were the procedures realistic?

No |__|__|__|__|__|__|__|__|__|__|__|__|__|__|__|__|__|__|__|__| Yes

Any Specific Problems?

At what level could you cope with the traffic in each of the following conditions, as a regular working level?

(100% = present day 95%   200% = twice present day 95%)

Weekday Morning One Sector

|__|__|__|__|__|__|__|__|__|__|__|__|__|__|__|__|__|__|__|__|

| .   .   .   .   !   .   .   .   .   . |

100   150   200

Weekday Afternoon One Sector

|__|__|__|__|__|__|__|__|__|__|__|__|__|__|__|__|__|__|__|__|

| .   .   .   .   !   .   .   .   .   . |

100   150   200

Weekend Morning One Sector

|__|__|__|__|__|__|__|__|__|__|__|__|__|__|__|__|__|__|__|__|

| .   .   .   .   !   .   .   .   .   . |

100   150   200

Weekend Afternoon One Sector

|__|__|__| |__|__|__|__|__|__|__|__|__|__|__|__|__|__|__|__|

| .   .   .   .   !   .   .   .   .   . |

100   150   200

Any Specific Problems?

At what level could you cope with the traffic in each of the following conditions, as a regular working level?

(100% = present day 95%   200% = twice present day 95%)

Weekday Morning Two Sector

|__|__|__|__|__|__|__|__|__|__|__|__|__|__|__|__|__|__|__|__|

| .   .   .   .   !   .   .   .   .   . |
100   150   200

Weekday Afternoon Two Sector

|__|__|__|__|__|__|__|__|__|__|__|__|__|__|__|__|__|__|__|__|

| .   .   .   .   !   .   .   .   .   . |
100   150   200

Weekend Morning Two Sector

|__|__|__|__|__|__|__|__|__|__|__|__|__|__|__|__|__|__|__|__|

| .   .   .   .   !   .   .   .   .   . |
100   150   200

Weekend Afternoon Two Sector

|__|__|__|__|__|__|__|__|__|__|__|__|__|__|__|__|__|__|__|__|

| .   .   .   .   !   .   .   .   .   . |
100   150   200

Any Specific Problems?

At what level could you cope with the traffic in each of the following conditions, as a regular working level?

(100% = present day 95%   200% = twice present day 95%)

Weekday Morning Two Sector

|__|__|__|__|__|__|__|__|__|__|__|__|__|__|__|__|__|__|__|__|

| .   .   .   .   !   .   .   .   . |

100   150   200

Weekday Afternoon Two Sector

|__|__|__|__|__|__|__|__|__|__|__|__|__|__|__|__|__|__|__|__|

| .   .   .   .   !   .   .   .   . |

100   150   200

Weekend Morning Two Sector

|__|__|__|__|__|__|__|__|__|__|__|__|__|__|__|__|__|__|__|__|

| .   .   .   .   !   .   .   .   . |

100   150   200

Weekend Afternoon Two Sector

| |__|__|__|__|__|__|__|__|__|__|__|__|__|__|__|__|__|__|__|

| .   .   .   .   !   .   .   .   . |

100   150   200

Any Specific Problems?

Are there any general or other problems or points you would like to make?

On behalf of the staff of the EREWHONCONTROL Experimental Centre, we thank you for your cooperation during this simulation. Your comments and suggestions will be seriously considered, and incorporated in the final report.

Operator Characteristics Questionnaire

These questions are asked only to investigate the possibility that some of the factors involved may affect performance. Answers to these questions are strictly confidential. Individual responses will **never** be disclosed in an identifiable form or published in either academic or business reports. The identity number marked on the questionnaire is the only link to recorded data. Please **do not write or sign your name** anywhere on this questionnaire. After the data has been transcribed into a secure computer data bank, the questionnaires will be destroyed. If you do not wish to take part in this survey, do not return the questionnaire, or leave it blank. Please circle one box in each line. You may leave blank any line you do not wish to answer:

Gender   Male   Female   Other

Age Range   20-24 25-29 30-34 35-39 40-44 45-49 50-54 55-59 60-64

Height   __ Ft __ inches   /   -.—Metres

Weight   -- Stone – Pounds (UK) /   ___ Pounds (USA) /   __._ Kg (Metric)

Education   Primary   Secondary   College or   Post-
(Highest Level)   Equivalent   Graduate

Relevant Experience
General   None   Less than   Less than   Less than   Less than   More Than
                   One month   one year   five years ten years   ten years

Specific   None   Less than   Less than   Less than   Less than   More Than
                   One month   one year   five years ten years   ten years

Most   Today Yester-   Last   Last   Last   None
Recent   day   Week   Month   Year

Have you discussed this experiment with previous participants? Yes/No
Computer Experience
How often do you use the following items?
ATM (Cash Point)   Never Daily   Weekly   Monthly Yearly
Computer   Never Daily   Weekly   Monthly Yearly
Keyboard   Never Daily   Weekly   Monthly Yearly
Mouse   Never Daily   Weekly   Monthly Yearly
Computer Games   Never Daily   Weekly   Monthly Yearly
Internet   Never Daily   Weekly   Monthly Yearly

What language do you normally think in?
What language did you think in during this experiment?
Do you have any disabilities that might affect your performance in this experiment?

Identity Number _____   **(DO NOT WRITE YOUR NAME)**

Chapter 12 – Running

After weeks or months of discussion, planning and preparation, the day will finally arrive when the simulation will run. This is the point when the choice of simulation leader has its effect. A huge and very expensive machine is ready to go, and, more important, so are the teams of controllers who play on this instrument like an orchestra.

Welcome

The first morning of a simulation is usually devoted to 'familiarisation'. This involves a welcome address from the Director, or some other suitable bigwig. If you have a publicity film, this is a good time to show it. A tour of the building should include the locations of lavatories, the canteen and any relevant offices. It should end up with a visit to the actual simulation room and an introduction to the simulator pilots and feed controllers. Participants can then be introduced to the equipment, and the layout of the simulator explained. They can be introduced to ISA (or SWAT if that is being used), NASA-TLX and any other measures. At this point, participants are issued with a card giving their name and an unique number (from 1 to 20). This is the identifying number used on questionnaires – actual identity is never written elsewhere.

Training

The first stage of running is invariably a training period. It is essential that training exercises should be taken as seriously as 'measured' exercises. Not only do the participants have to become familiar with equipment, which will inevitably be slightly different from what they are used to, but the teams have to shake down.

Although it is usually necessary to use reduced quantities of actual traffic during training exercises, it is essential that each training exercise should be taken exactly as seriously as the measured exercises. This means that all the feed sectors should be properly manned, and that all the measurement staff should be in place and working. In principle at least the measurement staff should be fully trained and not need further training. In reality, they are usually rusty, and may not be familiar with new or recently introduced measurement methods. They will certainly not be familiar with the participants, and experience suggests that the right sort of familiarity can contribute significantly to the enthusiasm of participants. If ISA (see chapter 10) or other quick-look analysis is to be used it should be used during training exercises, not only to familiarise the participants with any extra activities, but to make sure that the technique can produce results within the required time limit.

Training Exercises

Monday 12 1400 *Liz; Right everyone – First exercise – Team A – Monday 12 1400 – Single Sector, Weekday Morning Traffic 50% Traffic –Should be a doddle!*

Dr Whom - ISA all Magenta – Sector 1 Planner Red – Better check Back in a minute.

Dr W; OK – He had the scale upside down. No problem.

1415 Dr W; Traffic building up. ISA all cyan – Sector 1 planner Green.

1430 Liz; What's the matter with Strelsau Approach? Bill, go and look.

Bill; DUMMKOPF - Warning coming up as red on red. Igor will fix it this evening.

1453 Dr W: Traffic steady. ISA all Green, except Hentzau TMA - Amber

1520 Dr W; Zenda Approach No ISA – I'll remind him.

1545 Liz; Right, people Run completed. See you in Conference room A for debriefing in 15 minutes time.

No special points – equipment 'rather different'.

Monday 12 1600 Bill;   Second exercise – Team B – Monday 12 1600 Single Sector 50% - should be easy. (Strelsau Approach – we have turned up the colour contrast for DUMMKOPF warnings – it will be adjusted this evening)

1605 Dr W; ISA running – all Magenta just now

1630 Dr W; All ISA green, East and West Planners Amber.

1645 Dr W; Planners are back to Green, East Exec up to Amber

1700 Dr W All Green

1730 Dr W: Strelsau Approach to Amber

1736 Dr W;Strelsau Approach Red

1748 Dr W; Strelsau Approach Amber

1750 Liz; Thank You Ladies and Gentlemen – Debrief in Conference Room A in five minutes.

Let's get it wound up, your transport awaits. Enjoy the evening. Sleep Well.

-- -- ---

De-Brief note: Executives used to foot pedal not hand switch for frequency – Grumpy to fix accordingly.

Control Room Simulation

Tuesday 13 0900 *Liz; Good morning all. Right everyone – Team A –Two Sector layout, Weekday Morning Traffic 80% Traffic.*

0905 Dr W; ISA running – all Magenta just now

0908 Liz; Did they fix the DUMMKOPF Red on Red?

Dr W; : I think so – we'll have to wait till it comes up. DUMMKOPF isn't active today.

0923 Liz; What's up with the planners – I'll go and check.

0927 Liz; They don't like the handover point for UT1 – Note to discuss in de-briefing

0930 Dr W; All ISA green, East and West Planners Amber.

0945 Dr W; Planners are back to Green, East Exec up to Amber

1000 Dr W All Green

1030 Dr W: Strelsau Approach to Amber

1036 Dr W;Strelsau Approach Red

1048 Dr W; Strelsau Approach Amber

1100 Liz; Thank You Ladies and Gentlemen – Debrief in Conference Room A in five minutes.

-- -- ---

De-Brief Note : UT1 handover is too close to top of climb from Hentzau. Move 10Nm east.

Tuesday 13 1100 *Bill; Good morning all. Right everyone – Team B –Two Sector layout, Weekday Morning Traffic 80% Traffic.*

1105 Dr W; ISA running – all Magenta

1130 Dr W; All ISA green, East and West Planners Amber.

1145 Dr W; Planners are back to Green, East Exec up to Amber

1200 Dr W All Green

1230 Dr W: Strelsau Approach to Amber

1236 Dr W;Strelsau Approach Red

1248 Dr W; Strelsau Approach Amber

1255 Bill; Thank You Ladies and Gentlemen – Debrief in Conference Room A in five minutes.

-- -- ---

Nothing to report

___

Control Room Simulation

Tuesday 13 1400 Liz; Good afternoon all. Right everyone – Team A –Two Sector layout, Weekend Afternoon Traffic 80% Traffic.

1406 Dr W; ISA running – all Magenta just now

1430 Dr W; All ISA green, East and West Planners Amber.

1445 Dr W; Planners are back to Green, East Exec up to Amber

1400 Dr W All Green

1430 Dr W: Strelsau Approach to Amber

1436 Dr W;Strelsau Approach Red

1448 Dr W; Strelsau Approach Amber

1550 Liz; Thank You Ladies and Gentlemen – Debrief in Conference Room A in five minutes.

Nothing to report

Tuesday 13 1600 Bill; Good afternoon all. Right everyone – Team B –Two Sector layout, Weekend Afternoon Traffic 80% Traffic.

1606 Dr W; ISA running – all Magenta just now

1630 Dr W; All ISA green, East and West Planners Amber.

1645 Dr W; Planners are back to Green, East Exec up to Amber

1700 Dr W All Green

1730 Dr W: Strelsau Approach to Amber

1736 Dr W;Strelsau Approach Red

1748 Dr W; Strelsau Approach Amber

1755 Bill; Thank You Ladies and Gentlemen – Debrief in Conference Room A in five minutes

----

Liz; Looks good. They're not fully trained, but close enough to go to measured exercises. Agreed? Bill, Dr Whom: Agreed.

Liz; What was the problem in Strelsau?

Dr W; They thought they had two flights with the same number – poor printing on the strips. We are going through the samples to clear them up.

Control Room Simulation

Measured Exercises

Wednesday 14 0900 *Liz; Good morning all. Right everyone – Team B –Single Sector layout, Weekday Morning 100% Traffic.*

Lizl; Hang on – where are the Hentzau and Strelsau people? Bill, get on the blower – Dr W hold the simulation.

0920: Bill; Silly B&&&&&&s thought it was A team first – on their way in – should be here by 1030. Can we run an A team exercise?

Dr W; We'll have to cancel this one – it will take a good 20 minutes to set up – we'll run about an hour late – and I don't think all the A team are here anyway.

Liz; %%%%%% and &&&&&&&!

0935 Liz; All right people – stand down. See you – ALL OF YOU - 1400 hours.

Wednesday 14 1100 *Liz; Good morning all. Right everyone – Team A –Single Sector layout, Weekday Morning   100% Traffic.*

1109 Dr W; ISA – all magenta

1130 Dr W; All ISA Green Area Planner Amber

1133 Dr W; Strelsau Approach Amber

1136 Dr W; Strelsau Approach Red

1142 Dr W; Strelsau Approach Still Red

1151 Dr W; Strelsau Approach down to Amber

1154 Dr W; Everyone else Green

1200 Dr W; what's the matter now? Strelsau Approach in trouble

1200 Liz; tell them to disengage DUMMKOPF – sort it out afterwards.

1203 Dr W; Area Executive Red?

1206 Liz: Area Planner is helping Executive – find out afterwards what went wrong.

1224 Dr W; Area Executive to Amber

1230 Dr W; Area Exec Green – Hentzau Approach Red

1236 Liz; Hentzau approach sorted.

1254 Liz; Thank you Ladies and Gentlemen – Debrief in Conference Room A

-- -- ---

Strelsau Approach: We are getting warnings of conflict when there isn't any other flight within twenty miles – Digger says the sods are blinking on his display – WTHIH?[2]

Strelsau TMA : Too right Mate. We've got ghosts all over.

Area Planner : We're trying to cope but we have to redo everything two or three times. It is just not on.

Hentzau TMA : we are getting crossed wires all the time from Area.

Liz; Igor, you had better check on all these reports. You can use the replay facility over the weekend – I'll sort out someone to work with you – we'll suspend using DUMMKOPF until Monday.

---

[2] "What the hell is happening" – common aviation comment on automated systems.

Hentzau TMA : will that sort us out?

Liz : We'll have to suck it and see.

Liz : We had better re-run this one – Don't want team B to feel they are the only ones to come in on Friday. They have not got any junket organised for Friday, have they?

Wednesday 14 1400 *Liz; Good Afternoon everyone – Team B –Single Sector layout, Weekend Afternoon   100% Traffic. We won't be using DUMMKOPF for the moment, it seems to have some bugs. (Derisive cries from the floor.)*

1405 Dr W; ISA running – all Magenta just now

1430 Dr W; All ISA green, East and West Planners Amber.

1445 Dr W; Planners are back to Green, East Exec up to Amber

1400 Dr W All Green

1430 Dr W: Strelsau Approach to Amber

1436 Dr W;Strelsau Approach Red

1448 Dr W; Strelsau Approach Amber

1550 Liz; Thank You Ladies and Gentlemen – Debrief in Conference Room A in five minutes.

--- -- ---

DeBrief note : No major problems – familiarisation still going on

_____

Control Room Simulation

Wednesday 14 1600 *Bill; Good Afternoon everyone – Team A –Single Sector layout, Weekend Afternoon   100% Traffic. DUMMKOPF is off for the time being.*

Voice from the floor: I should f----ing think so!

(general agreement)

1609 Dr W; All running Magenta.

1630 Dr W; all running Green.

1635 Dr W; Strelsau TMA amber.

1700 Dr W; Strelsau TMA back to Green.

1730 Bill; Thank you ladies and gentlemen – debrief in Conference Room A in ten minutes.

-- -- ---

Bill; what was the problem in Strelsau TMA?.

Strelsau TMA : Nothing serious, computer seemed to be a bit slow.

Bill; I'll get Grumpy to look into it.

Strelsau TMA: Do we have to have that bloody heart rate monitor?

Bill; I'll see if we can get rid of it. Any other business?

See you tomorrow – 0900 prompt – don't miss the bus.

Thursday 15 0900 Bill; Good morning Hope you slept well – Igor didn't (Cheers) Team A - Two Sector layout, Weekday Afternoon 100% Traffic.

0905 Dr W; ISA running – all Magenta just now

0930 Dr W; All ISA green, East and West Planners Amber.

0945 Dr W; Planners are back to Green, East Exec up to Amber

1000 Dr W All Green

1015 Dr W: Strelsau Approach to Amber

1030 Dr W;Strelsau Approach Red

1042 Dr W; Strelsau Approach Amber

1045 Bill; Thank You Ladies and Gentlemen – Debrief in Conference Room A in five minutes.

Bill; Bit of a late peak in Strelsau – sorry about that.

Strelsau TMA : I think I fumbled that a bit – put Flavia into a bit of a squeeze.

Hentzau approach : Not the first time – OUCH! What was that for?

Bill; Any action needed?

Hentzau approach : Sorry, Flavia, didn't mean anything.

Strelsau : Just teething troubles – we can cope.

Hentzau approach; Not like that – I was teasing Hans - I didn't think

Strelsau approach; You never do! Men!

Bill; Right - See you again at 1400 – get your lunch in early - they are making Kartoffelsuppe!'

_____

Control Room Simulation

Thursday 15 1100 *Liz; Good morning Hope you slept well – Igor didn't (Cheers) Team B - Two Sector layout, Weekday Afternoon 100% Traffic.*

1106 Dr W; ISA – all magenta

1130 Dr W; All ISA Green East Sector Planner Amber

1133 Dr W; Strelsau Approach Amber

1136 Dr W; Strelsau Approach Green

1200 Dr W; All green – going like clockwork

1218 Dr W; East Sector Exec Amber

1219 Dr W; West Sector Exec Amber

1221 Dr W; East Sector Exec Red

1222 Liz; I'm going to look over the execs' shoulders – back in a minute.

1230 Dr W; East Sector Exec Amber

1248 Dr W ; East and West sectors Exec Green

1251 Dr W. All winding down Planners Magenta

1254 Lizl; Thank you Ladies and Gentlemen – Debrief in Conference Room A

No significant problems

Thursday 15 1400 Bill; Good Afternoon Team A - Two Sector layout, Weekend Morning 100% Traffic. Sorry to tell you, we'll have to rerun Wednesday morning tomorrow morning 0900 DON'T BE LATE! At the moment it looks as if Friday afternoon should be free.

1405 Dr W; ISA running – all Magenta just now

1430 Dr W; All ISA green, East and West Planners Amber.

1445 Dr W; Planners are back to Green, East Exec up to Amber

1400 Dr W All Green

1430 Dr W: Hentzau Approach to Amber

1436 Dr W; Hentzau Approach Red

1440 Dr W; Hentzau Approach Amber

1448 Dr W; Hentzau Approach Green

1550 Bill; Thank You Ladies and Gentlemen – Debrief in Conference Room A in five minutes.

-- -- ---

No significant problems

Control Room Simulation

Thursday 15 1600 *Liz; Good Afternoon Team B - Two Sector layout, Weekend Morning 100% Traffic. We'll have to rerun yesterday morning's exercise tomorrow morning 1100, but you should be free for the weekend after that – unless something else goes wrong.*

1606 Dr W; ISA running – all Magenta just now

1630 Dr W; All ISA green, East and West Planners Amber.

1645 Dr W; Planners are back to Green

1700 Dr W All Green

1730 Dr W: Hentzau Approach to Amber

1736 Dr W; Hentzau Approach Green

1755 Liz; Thank You Ladies and Gentlemen – Debrief in Conference Room A in five minutes

-- -- ---

No significant problems

Friday 16 0900 *Bill: Good morning all. Right everyone – Team B –Single Sector layout, Weekday Morning   100% Traffic. This is a repeat of Wednesday morning that did not happen – you know why. (Groans from the floor)*

0905 Dr W; ISA running – all Magenta just now

0930 Dr W; All ISA green, East and West Planners Amber.

0945 Dr W; Planners are back to Green,

1000 Dr W All Green

1015 Dr W: Strelsau Approach to Amber

1020 Dr W;Strelsau Approach Green – All Green

1045 Bill; Thank You Ladies and Gentlemen – Debrief in Conference Room A in five minutes.

Strelsau approach mildly stressed.

No significant problems

Friday 16 1100 *Liz; Good morning all. Right everyone – Team A –Single Sector layout, Weekday Morning 100% Traffic. This is a rerun of Wednesday morning, without DUMMKOPF.(Cheers from the floor)*

1109 Dr W; ISA – all magenta

1130 Dr W; All ISA Green Area Planner Amber

1133 Dr W; Strelsau Approach Amber

1136 Dr W; Strelsau Approach Red

1142 Dr W; Strelsau Approach Still Red

1151 Dr W; Strelsau Approach down to Amber

1154 Dr W; Everyone else Green

1213 Dr W; All Green

1245 Liz; Thank You Ladies and Gentlemen – Debrief in Conference Room A in five minutes.

-- -- ---

Strelsau Approach loaded at peak

No significant problems

All controllers stood down until Monday 19 0900

Igor and Grumpy will come in tomorrow and, if necessary Sunday, to sort out the interface problems with DUMMKOPF.

Monday 19 0900 Bill; Good morning Hope you had a good weekend Team B - One Sector layout, Weekday Morning 125% Traffic.

0905 Dr W; ISA running – all Magenta just now

0930 Dr W; All ISA green, East and West Planners Amber.

0945 Dr W; East Planner are back to Green, East Exec up to Amber

1000 Dr W; West Planner Green

1015 Dr W: Strelsau Approach to Amber

1030 Dr W; Strelsau Approach Red

1039 Dr W; East Exec to Green

1042 Dr W; Strelsau Approach Amber

1045 Bill; Thank You Ladies and Gentlemen – Debrief in Conference Room A in five minutes.

-- -- ---

No significant problems

Control Room Simulation

Monday 19 1100 Liz; Good morning Hope you had a good weekend Team A - One Sector layout, Weekday Morning 125% Traffic.

1106 Dr W; ISA – all magenta

1130 Dr W; All ISA Green Sector Planner Amber

1133 Dr W; Strelsau Approach Amber

1136 Dr W; Strelsau Approach Green

1200 Dr W; All green – going like clockwork

1218 Dr W; East Sector Exec Amber

1219 Dr W; West Sector Exec Amber

1221 Dr W; East Sector Exec Red

1230 Dr W; East Sector Exec Amber

1248 Dr W ; East and West sectors Exec Green

1251 Dr W. All winding down Planners Magenta

1254 Lizl; Thank you Ladies and Gentlemen – Debrief in Conference Room A

East Sector Executive mildly overloaded – Training Problem. No other significant problems

Monday 19 1400 *Liz; Good Afternoon Team B - One Sector layout, Weekend Afternoon 125% Traffic.*

1405 Dr W; ISA running – all Magenta just now

1430 Dr W; All ISA green, Sector Planner Amber.

1445 Dr W; Planner is back to Green, Exec up to Amber

1400 Dr W All Green

1430 Dr W: Hentzau Approach to Amber

1436 Dr W; Zenda Approach Amber

1500 Dr W; Hentzau and Zenda Approaches both Red

1515 Dr W; Hentzau and Zenda to Amber

1539 Dr W; Hentzau and Zenda Approaches to Green

1550 Liz; Thank You Ladies and Gentlemen – Debrief in Conference Room A in five minutes.

-- -- ---

No significant problems Hentzau and Zenda still training effects.

No Heart Rate readings from Strelsau – Heart Rate monitor "missing" ?

Control Room Simulation

Monday 19 1600 Bill; Good Afternoon Team A - One Sector layout, Weekend Afternoon 125% Traffic.

1606 Dr W; ISA running – all Magenta just now

1630 Dr W; All ISA green, Sector Planner Amber.

1645 Dr W; Planner back to Green,

1700 Dr W All Green

1730 Dr W: Hentzau Approach to Amber

1736 Dr W; Hentzau Approach Red, Zenda Approach Red

1748 Dr W; Zenda Approach Amber

1750 Dr W; Hentzau Approach Green

1755 Bill; Thank You Ladies and Gentlemen – Debrief in Conference Room A in five minutes

----

No significant problems –

Hentzau and Zenda Approach still not fully up to scratch.

Heart Rate Monitor still "Missing"

Tuesday 20 0900 *Liz; Good Morning Team A - Two Sector layout, Weekday Afternoon 125% Traffic. DUMMKOPF running*

0906 Dr W; ISA running – all Magenta just now

0930 Dr W; All ISA green, East and West Planners Amber.

0945 Dr W; Planners are back to Green, East Exec up to Amber

1001 Strelsau Approach; I've gone dead. The bloody system has stopped.

What does ERROR 2013 have to do with anything?

1003 Dr W; whole system has stopped.   Can you fix it – Grumpy? Igor?

1005 Liz; Sorry, we have a problem. We have to abort this run. See you 1300

Igor, Grumpy, Bill, Dr W - Conference Room A in five minutes.

Strelzau Approach; Why does your system stop on a single glitch? We would be in deep trouble if our system packed up like this.

Grumpy: As soon as they double our software budget, we will be able to put in a full fall-back system. Until then all we lose is one or two exercises per year – we are not putting lives at risk.

Grumpy; PA003 came up with a time in sector of minus 712 minutes. The system tried to cancel it about twelve hours before it arrived. It couldn't find it, but cancelled the lead record of the stack. Crash! Tinkle!

Dr W.; Igor, when you set times, where does DUMMKOPF get Real-Time values from?

Igor; I think it uses the standard TIMEX function. We have never had any problems with it.

Grumpy; Hang on – I think I saw something about that one .Back in a moment.

Grumpy; Here we are. Timex gives time in seconds from midnight – local time.

Local time – should be set to appropriate time zone – otherwise GMT is used. We are ten hours off GMT.

Igor; We – I – didn't think of that. I'll put in a reset and check for time zone.

Liz: Igor and Grumpy- do a check run this evening. We won't use DUMMKOPF for the 11 00 exercise – same time range. We'll risk it for the afternoon exercises.

Control Room Simulation

Tuesday 20 1100 Bill; Good Morning Team B – Two Sector layout, Weekday Afternoon 125% Traffic DUMMKOPF NOT running – hiccup in last exercise.

1106 Dr W; ISA – all magenta

1130 Dr W; All ISA Green East Sector Planner Amber

1133 Dr W; Strelsau Approach Amber

1136 Dr W; Strelsau Approach Green

1200 Dr W; All green – going like clockwork

1218 Dr W; East Sector Exec Amber

1219 Dr W; West Sector Exec Amber

1221 Dr W; East Sector Exec Red

1230 Dr W; East Sector Exec Amber

1248 Dr W ; East and West sectors Exec Green

1251 Dr W. All winding down Planners Magenta

1254 Bill; Thank you Ladies and Gentlemen – Debrief in Conference Room A

-- -- ---

No significant problems

Tuesday 20 1400 *Liz; Good Morning Team A - Two Sector layout, Weekend Afternoon 125% Traffic. DUMMKOPF Running*

1405 Dr W; ISA running – all Magenta just now

1430 Dr W; All ISA green, East and West Planners Amber.

1445 Dr W; Planners are back to Green, East Exec up to Amber

1400 Dr W All Green

1430 Dr W: Strelsau Approach to Amber

1436 Dr W;Strelsau Approach Red

1448 Dr W; Strelsau Approach Amber

1550 Liz; Thank You Ladies and Gentlemen – Debrief in Conference Room A in five minutes.

-- -- ---

No significant problems

Control Room Simulation

Tuesday 20 1600 Liz; Good Afternoon Team B – Two Sector layout, Weekend Afternoon 125% Traffic, .DUMMKOPF Running

1606 Dr W; ISA running – all Magenta just now

1630 Dr W; All ISA green, East and West Planners Amber.

1645 Dr W; Planners are back to Green, East Exec up to Amber

1700 Dr W All Green

1730 Dr W: Strelsau Approach to Amber

1736 Dr W; Strelsau Approach Red

1748 Dr W; Strelsau Approach Amber

1755 Liz; Thank You Ladies and Gentlemen – Debrief in Conference Room A in five minutes

-- -- ---

No significant problems

Wednesday 21 0900 *Bill; Good morning Team B - One Sector layout, Weekday Morning 150% Traffic0905 Dr W; ISA running – all Magenta just now*

0915 Dr W; All TMA green, Sector Planner Amber.

0920 Dr W; Planner is Red, Sector Exec up to Amber

0927 Dr W; All TMAs Amber, Sector Planner and Exec Red.

0945 Dr W: Strelsau Approach to Red

0954 Dr W; Hentzau Approach Red

1010 Dr W; Planner Amber

1024 Dr W; Planner Red

1030 Dr W; Hentzau Approach Amber

1035 Dr W; Planner Amber

1038 Bill; All right – Stand down – that was a tough one. Take a breather and debrief in Conference Room A in five minutes.

Bill; Well, this is clearly too much. We rather expected that. Not to worry, it's the system that fails, not you.

Chill out till lunch – I gather they are making Kartoffelsuppe.

Ad try not to put the wind up Team B – let them find out for themselves.

Control Room Simulation

Wednesday 21 1100 Bill; Good morning Team A - One Sector layout, Weekday Morning 150% Traffic. This may well be too much. 1106 Dr W; ISA – all magenta

1108 Dr W; All ISA Green Sector Planner Amber

1110 Dr W; Strelsau Approach Amber

1116 Dr W; Strelsau Approach Green

1120 Dr W; Planner Amber, Executive Amber

1125 Dr W; Planner Red

1136 Dr W; Strelsau Approach Amber

1142 Dr W; Strelsau Approach Red

1148 Dr W; Planner Amber, Executive Red

1200 Dr W; Executive Amber

1218 Dr W; Strelsau Approach Amber

1227 Dr W; Strelsau Approach Red

1235 Bill; All right Gentlemen, Ladies, Stand down, Team A had much the same, by the way.

1240 Bill; Thank you Ladies and Gentlemen – Debrief in Conference Room A

Bill; This one was almost an overload – we'll have to try with the weekend traffic, and then with two sectors, which should be just possible.

Sector Exec; We could have done better if I'd had the strips before the traffic turned up.

Sector Planner; Sorry, Fritz, I had to get the levels right, and there just wasn't enough time.

Bill; Don't worry – this is where we find out – better now than in real life.

In the meantime there is KartoffelSuppe for lunch.

Wednesday 21 1400 Liz; *Good Afternoon Team B - One Sector layout, Weekend Afternoon 150% Traffic1405 Dr W; ISA running – all Magenta just now*

1415 Dr W; All ISA green, Sector Planner Amber.

1420 Dr W; Planner is Red, Sector Exec up to Amber

1427 Dr W; All TMAs Amber. Exec Red

1435 Dr W: Strelsau Approach to Red

1437 Dr W; Hentzau Approach Red

1451 Dr W; Hentzau Approach Amber

1457 Dr W; Planner Amber

1512 Dr W; Strelsau Approach Amber

1518 Dr W; Exec Amber

1524 Dr W; Hentzau Approach Amber

1527 Dr W; Planner Red

1530 Liz; Right - Stand down. Take a breather and debrief in Conference Room A in ten minutes.

-- -- ---

Sector Planner: I'm sorry, there was just too much coming in. I had to hang on to the strips – Sorry.

Liz; No problem, no one could have coped with all that lot any better.

Control Room Simulation

Wednesday 21 1600 Liz; Good Afternoon Team A - One Sector layout, Weekend Afternoon 150% Traffic 1606 Dr W; ISA running – all Magenta just now

1609 Dr W; Planner Cyan

1610 Dr W; Planner to Red, Sector Exec Amber

1615 Dr W; All ISA Amber, Sector Planner and Exec Red

1625 Dr W; TmAs ISA Green, Hentzau Approach Amber,

1630 Dr W; Sector Planner Red Exec Amber

1645 Dr W; All ISA Amber, Sector Planner Amber, Exec Red

1700 Dr W; Hentzau Approach Red

1715 Dr W; Hentzau Approach Amber, Sector Planner and Exec Red

1718 Dr W; Sector Planner Amber

1721 Dr W; Sector Planner Red Exec still Red

1730 Dr W; Exec Red

1735 Liz; All Controllers Stand down. Exercise completed.

1739 Liz; Thank You Ladies and Gentlemen – Debrief in Conference Room A in five minutes

---

Liz; Well, That was tough. Frankly, I did not think you could make it. Congratulations.

Controllers; We could not keep that up every day. Sorry about the mess over HES

Liz; That should not have happened in the first place – it was practically impossible to sort out by the time you got it.

Liz; We have those bloody visitors tomorrow – please God we don't have another one like this.

Thursday 22 – visit of Party from Ruritania.

0900 Centre driver picks up Marshal Strakenz and Colonel Sapt from the Ruritanian Army, Captain Von Tarlenheim from the Ruritanian Air Force and Monsieur Ronde de Cuir from the Ruritanian Air Ministry. In addition, R. Hentzau from Ruritania Control joins the party. Centre driver informs centre by cellphone of addition.

1000 Party arrives at Centre. Greeted by director's secretary – offered coffee and 'Hundekuchen' – Ushered to Conference Room B.

1015 Director addresses party.

Dir: Good Morning, Gentlemen

"*As you know, EREWHONTROL was set up in 1968 to organise and upgrade Air Traffic Services in Western Erewhon, The Experimental Centre was set up in 1970 to assist in the development of ATC centres in the member countries, with a view to introducing cross-national centres. For example, the outstanding centre at Blefuscu has coped in the remarkable way you will know with the peculiar problems of coordination between Lilliput and Brobdignag. Blah, Blah, Blah.*"

****

"*In conclusion, may I welcome you to the Centre, and may we show you the specially prepared video, which explains the role of Real-Time Simulation in Air Traffic Control.*

We will then show you around the centre, and Squadron Leader Bennet will show you the simulation now under way. Since this is a measured exercise, we would ask you, please, not to talk to, or distract the controllers during the simulation.

We have arranged lunch in the Senior Staff Room at 12.30, and the controllers have invited you to join them after lunch at about 1400 in the Participants' Room – a rare honour, by the way.

We have arranged transport for 1500 hours to the Airport. (I understand that Mr. Hentzau will be staying for some days.)

I regret that I have other pressing matters to attend to, but I hope to join you at lunch."

1100 Visitors viewed "Erewhontrol, Past, Present and Future" and

"*The Role of the EREWHONTROL Experimental Centre in the development of Safe, efficient and economic ATC*"

S/L Bennet joined them at 1115 and the party discussed aspects of Real-Time simulation arising from the Director's address and the videos.

Control Room Simulation

M. Ronde de Cuir enquired particularly about the financial implications of running the Real-Time Simulator. In reply S/L Bennet provided him with the current financial statement, a paper comparing costs (favourably) with certain other centres, and finished by reminding him of the saying "If you think safety costs money, try paying for an accident."

Marshal Strakenz asked about the effect of civil traffic on military activity. Apart from a prohibited zone around Martburg Training area and a minimum height restriction over seven designated security zones, there was very little mutual inconvenience.

Captain Von Tarlenheim added that most flying training took place around Martburg, or in the dedicated combat training area in Nephelocccygia. Military ATC received the same input as civil 'plus certain other information', and was responsible for keeping military traffic separated from civil. Some military transport flights used civil airways, but conformed to civil ATC when doing so.

The party then were shown over the Experimental Centre, viewing the historical and development gallery, the data analysis centre and other major parts of the centre. They then visited the Simulation Room, where they observed the current Ruritania I simulation, where a two-sector weekday afternoon sample with 150% of peak traffic was running, with DUMMKOPF in action. Igor Igorovitch explained the function of DUMMKOPF and pointed out how it assisted the Strelsau approach controller.

After termination of the exercise at 12.54, Dfr Hildegard Von Bingen and Dfr Flavia Elphburg joined the visitors at lunch in the Director's Dining Room.

FE; I have always wondered how you knew to hold off the flank attack at Munchausen – The strain must have been terrific.

Marshal Strakenz; Actually, I fell asleep, and forgot to send the order. When I woke up, I sent it immediately, which meant they got to the bridge as the enemy was half-way over and our boys caught them on the hop. If I had sent it when I meant to, we would have been trapped and wiped out. God looks after fools, drunks and cavalry officers, so I had a triple insurance.

FE; I suspect you are being modest, Marshal.

MS; I have a lot to be modest about.

FE; You borrowed that from Churchill – about Attlee.

MS; I wondered where it came from.

After lunch, the visitors were invited to join the off-duty controllers in the controllers' lounge, where they were able to discuss the simulation, about which the controllers expressed satisfaction, and DUMMKOPF about which they did not.

Finally, the visiting party were presented with documentation folders and copies of the EREWHONTROL presentation volume, keyrings and tiepins and signed the Visitors' Book. The Director, S/L Bennett, Dfr Von Bingen and Dfr Elphburg, then saw them off to the airport, except for Dhr. Rupert Hentzau, who remained on simulation business.

Control Room Simulation

Thursday 22 0900 Liz; Good Morning Team A - Two Sector layout, Weekday Afternoon 150% Traffic. DUMMKOPF Running

0905 Dr W; ISA running – all Magenta just now

0930 Dr W; All ISA green, East and West Planners Amber.

0945 Dr W; Planners are Red, East Exec up to Amber

1000 Dr W Planners Amber, All others Amber

1015 Dr W: Strelsau Approach to Amber

1030 Dr W; Hentzau Approach Amber, Strelsau Red

1042 Dr W; Zenda Approach Amber

1043 Dr W; Planners Green

1045 Liz; Thank You Ladies and Gentlemen – Debrief in Conference Room A in five minutes.

--- -- ---

Liz; Congratulations – I mean that- that was about as heavy a load as anyone could cope with using this equipment.

RUE Planner: I don't think I could cope with that level on a daily basis.

Strelsau Approach; I'm not happy about DUMMKOPF – I have to think twice all the time – what should I do? Why doesn't DUMMKOPF agree? And it is still doing some peculiar things.

Thursday 22 1100 *Liz; Good Morning Team B – Two Sector layout, Weekday Afternoon 150% Traffic. DUMMKOPF Running,*

1102 Liz; Bill, will you take over; I have to go and herd sheep – two pongos, a fly-boy and a paper-pusher from Strelsau on a joy ride. See you…

1106 Dr W; ISA – all magenta

1110 Dr W; All ISA Green East Sector Planner Amber

1113 Dr W; Strelsau Approach Amber

1114 Dr W; Hentzau Approach Green

1115 Dr W; East Planner Red

1117 Dr W; All Amber

1204 Dr W; Strelsau Approach Red

1219 Dr W; West Sector Exec Red

1221 Dr W; East Sector Exec Red

1225 Dr W; Planners Amber

1230 Dr W; East Sector Exec Amber

1248 Dr W; Approaches Amber

1248 Dr W; East and West sectors Exec Amber

1251 Dr W; All winding down Planners Magenta, others Green

1254 Bill; Thank you Ladies and Gentlemen – Debrief in Conference Room A

-- -- ---

Liz; Well, folks – that was a heavy one. Just as well that you made it. The visitors were very impressed.

RE Planner; What visitors?

Hentzau TMA; Didn't you notice them. Some Army Brass and a couple of others.

Strelsau Approach; Wasn't that old Strakenz? I've always wanted to meet him.

Liz; Now's your chance – You and Helga are invited to lunch with them and the director – It's on the house – not KartoffelSuppe.

OK, See you all 1600 hours – until then relax… You did a good job.

Control Room Simulation

Thursday 22 1400 *Liz; Good Afternoon Team A - Two Sector layout, Weekend Afternoon 150% Traffic. DUMMKOPF Running,*

1405 Dr W; ISA running – all Magenta just now

1410 Dr W; All ISA green, East and West Planners Amber.

1425 Dr W; Planners both Red, East Exec up to Amber

1429 Dr W; West Exec to Amber

1430 Dr W: Strelsau Approach to Amber, West Exec Red

1432 Dr W: Hentzau Approach to Amber

1435 Dr W: Strelsau Approach to Red

1437 Dr W: Zenda Approach to Amber

1440 Dr W: Hentzau and Zenda Approach to Red

1445 Dr W: East Planner Amber

1436 Dr W; Strelsau Approach Red

1448 Dr W; Strelsau Approach Amber

1453 Dr W; Strelsau Approach Amber

1500 Dr W; Strelsau Approach Amber

1515 Dr W; All Amber

1535 Dr. W; Planners Green

1540 Dr W; TMAs Green

1550 Liz; Thank You Ladies and Gentlemen – Debrief in Conference Room A in five minutes.

Liz; I'm impressed – you are really getting on top of it now, although you are obviously working hard. Strelzau Approach how was it?

Strelsau Approach; Pretty tough – I have to say I simply ignored DUMMKOPF – it seemed to be completely out of step. And it gets in the way. Do we still have to have it around our necks?

Liz; Afraid so. Will you try to tell Igor what you found wrong? Everyone else – relax till the bus comes. Let the others take the strain.

Thursday 22 1600 *Liz; Good Afternoon Team B – Two Sector layout, Weekend Afternoon 150% Traffic. DUMMKOPF Running,*

1605 Dr W; ISA running – all Magenta just now= East Planner Amber

1610 Dr W; All ISA Cyan, East and West Planners Amber.

1625 Dr W; Planners both Red, East Exec up to Amber, all others Green

1620 Dr W; West Exec to Amber

1630 Dr W: Strelsau Approach to Amber, West Exec Red

1632 Dr W: Hentzau Approach to Amber

1635 Dr W: Strelsau Approach to Red

1837 Dr W: Zenda Approach to Amber

1640 Dr W: Hentzau and Zenda Approach to Red

1645 Dr W: East Planner Amber

1648 Dr W; Strelsau Approach Amber

1653 Dr W; Hentzau Approach Amber

1700 Dr W; Zenda Approach Amber

1715 Dr W; All Amber

1735 Dr. W; Planners Green

1740 Dr W; TMAs Green

1650 Liz; Thank You Ladies and Gentlemen – Debrief in Conference Room A in five minutes.

Liz; Hard work, but you seem to be coping well. I told them you could handle it. Strelzau Approach how was it?

Strelsau Approach; Not so bad, but do we have to use DUMMKOPF. I don't claim to understand it, but it just doesn't seem to have the same priorities as we do.

Liz; Will you try to tell Igor what you found wrong? I'll run you back to your hotel afterwards. Everyone else – the bus is ready to go – have a pleasant evening.

Control Room Simulation

Friday 23 0900 *0900 Liz; Good Morning Team A - Two Sector layout, Weekday Afternoon 125% Traffic. DUMMKOPF running This exercise replaces Tuesday Morning.*

0903 Dr W; ISA running – all Magenta just now

0910 Dr W; All ISA cyan, East and West Planners Green

0918 Dr W;.All ISA Green

0920 Dr W; Planners Amber

0945 Dr W; Planners are back to Green, East Exec up to Amber

1000 Dr W All Green

1030 Dr W: Strelsau Approach to Amber

1036 Dr W;Strelsau Approach Red

1040 Dr W; Strelsau Approach Amber

1045 Dr w; All Green

1055 Liz; Thank You Ladies and Gentlemen – Debrief in Conference Room A in five minutes

-- -- ---

No significant problems.

Well, Ladies and gentlemen, that wraps up this week. The coach will leave after the next exercise, so just relax and brace yourselves for the week-end,

Friday 23 1100 Liz; Good Morning Team B - Two Sector layout, Weekend Morning 100% Traffic This is a repeat of last Thursday afternoon – some silly sod overwrote the records so we have to do it again. The traffic is still 100%, but it will be different from the previous time. This should be an easy one.

1103 Dr W; ISA running – all Magenta just now

1110 Dr W; All ISA cyan, East and West Planners Green

1118 Dr W; .All ISA Green

1120 Dr W; Planners Amber

1145 Dr W; Planners are back to Green, East Exec up to Amber

1200 Dr W All Green

1230 Dr W: Strelsau Approach to Amber

1236 Dr W; Strelsau Approach Red

1240 Dr W; Strelsau Approach Amber

1245 Dr w; All Green

1255 Liz; Thank You Ladies and Gentlemen – Debrief in Conference Room A in five minutes

----

No significant problems

Goodbye until Monday. Enjoy your weekend. Next week starts with some tough ones.

The coach is eager to go so do not miss it.

Control Room Simulation

26 0900 Bill; Good morning Hope you had a good weekend Team B - One Sector layout, Weekday Morning 175% Traffic. This is going to be tough.

0905 Dr W; ISA running – all Magenta , Planners Cyan

0910 ALL POWER OFF

0911 Liz: ALL PERSONNEL Follow the emergency lights, follow instructions from the people in High-Visibility jackets. Everyone report to the gatehouse and stay there until further orders. Fire Marshalls; ensure everyone has cleared the building.

0920 Liz; at least it isn't raining. No signs of fire – anyone know what has happened?

0935 Liz; All present and correct. Now all we can do is wait. You can go to your cars if you want to keep warm.

0932 Gate Guard; This man is from the building site over there. He says some fool cut the local power line with a digger. The emergency people are on the way.

0925 Liz; You can all return to the building now. Power is still off. Do NOT turn anything on or off. Power will be restored shortly. Bill, Grumpy, Dr W. a word, please.

0930 Liz; Well, that's the first exercise buggered. How long will it take to get the 1100 exercise going from cold?

Grumpy; From a cold start, about fifteen minutes. Can't do anything without power.

Bill; I have the paperwork ready - I'll pass the word for Team B to assemble at 10.45. We'll have to re-run this one on Friday morning.

Dr W; Hang on . Wouldn't it be better to move this one to Wednesday Morning, and put the Wednesday one to Friday. They will be back to 100% then and another 175% won't go down very well.

Liz; Yes – I see what you mean. O.K? Bill? Grumpy?

Bill; O.K.

Grumpy; Agreed.

Liz; Make it so, then. Let's get indoors.

Monday 26 1100 Bill; Good morning Hope you had a good weekend Team A. I'm sorry to say we haven't been having a good time this morning. As you have found out, we are having a power cut. It should be fixed for this afternoon. I know, all real centres have several alternative supplies, but the powers that be decided it was not economic for a non-essential installation. The canteen manager says he will have hot food on time if power comes back before 1130, otherwise it will be hot soup and assorted cold dishes. In any case, we can't run this morning. We are shuffling things around and we will be running two exercises on Friday morning. In the mean time you are free till 1600 hours, so enjoy yourselves.

11.10 Ad-hoc Meeting Liz, Bill, Dr W. Grumpy, Dopey, Grandad, Bashfull Bashfullson.

Liz; They swear they'll have power back in a few minutes. In the meantime, we have had to re-arrange this week's exercises. We are putting both this morning's exercises on to Wednesday morning, and putting Wednesday's on to Friday. If nothing else happens we should be OK.

Grandad: I'm a bit worried about this week's exercises, especially for the one sector organisation, I don't think they can handle it. We got too many late handovers with 150% last week, and the lads are not at all happy.

Liz; You are probably right – I'll tip you the wink if it is getting really hairy, and you can stop sending any further traffic. Make sure Jack the Lad and Wendy know as well. Where is Jack, by the way?

Grandad; Ah… Slight case of Monday Morning I think. He went off with Rupert Hentzau last Friday afternoon. We haven't seen either of them since.

Liz; And you were covering for them – I wouldn't expect anything else.

1130: POWER ON

1135 Liz; All safety checks OK – Put it on the Tannoy, Exercise for Team B at 1400, Team A at 1600

We'll do both morning exercises on Wednesday and both Wednesday morning on Friday Morning.

Dopey – I have to tell you, someone has overwritten last Thursday's records – we will have to run it again.

Liz; there goes Friday afternoon – dammit. Is that A or B?

Dopey; B I think – sorry, Liz I mean It *is* Team B.

Control Room Simulation

Monday 26 1400 *Liz; Good Afternoon Team B – I'm sorry to say we will have to run this morning's exercise on Friday morning, and in addition we have to rerun last Thursday afternoon on Friday afternoon – some silly sod lost the records. For now, however - One Sector layout, Weekend Afternoon 175% Traffic. This is likely to be very difficult. I will abort the exercise if it is clearly not working.*

1405 Dr W; ISA running – all Magenta just now

1415 Dr W; All ISA green, Planner Amber.

1420 Dr W All Approach Amber

1422 Liz; Grandad, hold all incoming traffic.

1428 Dr W; Planner is Red, Exec up to Amber

1430 Dr W: Strelsau Approach to Red

1436 Dr W; Hentzau Approach Red

1438 Dr W; Zenda Approach Amber

1440 Liz; All right- this clearly is not working. We will stop now. Thank You Ladies and Gentlemen – Debrief in Conference Room A in five minutes.

Sector Planner; That was simply too much. The frequency was jammed – even with Jim helping me I couldn't handle that lot.

Sector Exec; I was getting aircraft on the frequency while Hans was still holding on to the strips.

Approaches; We couldn't get clearances for departures, and bloody DUMMKOPF was doing its nut – definitely a nono!

Liz; Definitely a write-off, but I was impressed with how far you got. We will have to do a few more to see if any other samples come through better. Thank you for the skill and devotion. Go and have a drink and relax till the bus comes round. Sleep well – no-one could have done better.

Monday 26 1600 Bill; Good Afternoon Team A I have to tell you we will have to run this morning's exercise on Friday morning. However, today it is - One Sector layout, Weekend Afternoon 175% Traffic. As you will have heard from your colleagues, this is practically impossible, but we will see how far we can get.

1606 Dr W; ISA running – all Magenta

1612 Dr W; All ISA green, Planner Amber.

1615 Dr W; Planner Red, Exec up to Amber

1620 Dr W All Approach to Amber, Exec Red

1625 Bill; Grandad, Hold all incoming traffic

1630 Dr W; All Red

1630 Bill; Thank You Ladies and Gentlemen – This one is clearly much too much. We rather expected it to be, so don't blame yourselves. You are testing the system we are not testing you. Debrief in Conference Room A in five minutes

--- -- ---

Sector Exec; There just was not enough time to get the picture. Willie couldn't get the clearances through in time. A complete f****-up.

(General Agreement)

Tuesday 27 0845 *Liz; Ah, There you are Jack.*

Jack; Sorry, Liz, I was not very well yesterday. I should have phoned in, but I could not stand the noise of the dialling tone.

Liz; Have you seen Rupert Hentzau since the weekend. He seems to have had a touch of the same illness.

Jack. No, We had quite a bash on Saturday evening, I got back about 6 in the morning. Rupert was still out -, his key was still there. I think he met a 'friend'. He was carrying on as usual with one of the waitresses at the Schweinerie Haus. I saw them whispering about something.

Liz; I'll get Wendy to phone him. No, you had better do it – let me know when he proposes to honour us with his presence.

0855 Jack; I've phoned the Puteairelle and they have checked – he has not been back – his bed has not been slept in – "single or double" as they put it.

Liz; Well. If he has not turned up by tomorrow, we had better start making noises. For now, I have better things to do.

Jack; I'll try the Schweinerie – see if anyone knows anything – I think the girl was called Vlada.

Tuesday 27 0900 *Liz; Good Morning Team A - Two Sector layout, Weekday Afternoon 175% Traffic. DUMMKOPF Running,*

0909 Dr W; All ISA Cyan, East and West Planners Green

0910 D. W; ERROR 2013 – Bloody DUMMKOPF has stalled again.

0911 Liz; All right, stand down all. Don't go away, we may be able to re-start.

L:iz; Grumpy, can you re-set? Cut out DUMMKOPF.

Grumpy; I'm keeping a DUMMKOPF free version on hand for every exercise now. We will be ready to go in say 15 minutes.

Liz; All right, people, go and get a coffee. We will re-start at 0945.

Tuesday 27 0945 *Liz; Well, here we are again. Two Sector layout, Weekday Afternoon 175% traffic – And no DUMMKOPF (Sorry Igor)*

0950 Dr W; ISA all Magenta, Planners Green

0955 Dr W; Planners to Amber, East Exec up to Amber, West Exec to Red

1000 Dr W; Planners Red, Both Exec Amber, rest Green

1005 Dr W; Planners Red, Execs Red, Strelsau TMA Amber

1010 Liz; Grandad, hold incoming off traffic..

1015 Dr W; Zenda and Hentzau TMA amber

1016 Dr W: Strelsau Approach to Amber

1020 Dr W; Strelsau Approach Red

1022 Dr W; All Red

1025 Liz; Thank You Ladies and Gentlemen. I'm calling this one off, it clearly isn't working. – Debrief in Conference Room A in five minutes.

Planners; You didn't have to call it off Liz, we were just getting it under control?

Liz; I don't want to push you too hard, but you will have another chance this afternoon.

Liz; I'm sorry Igor, but we can't run DUMMKOPF with this traffic. Why don't you look at some of the recordings? They might help sort out the problems.

Control Room Simulation

Tuesday 27 1100 Liz; Good Morning Team B – Two Sector layout, Weekday Afternoon 175% Traffic. DUMMKOPF Suspended for now.

1106 Dr W; ISA – all magenta

1110 Dr W; All ISA Green, East Planner Amber

1113 Dr W; Both Planners Amber

1117 Dr W; Both Planners Red, Both Exec Amber

1120 Dr W; Planners Red, Execs Red, others Green

1124 Dr W; East Planner missed the ISA – make it Red

1130 Dr W; All approach Amber

1137 Dr W; West Exec Amber

1145 Dr W; East Planner Amber Exec Red

1200 Dr W; East Exec Amber

1208 Dr W; West Exec and Planner Amber

1215 Dr W; Strelsau Approach Red

1225 Dr W; Strelsau Approach Amber

1245 Dr W; All winding down Planners Magenta, Approach Green

1254 Liz; Thank you Ladies and Gentlemen – Debrief in ten minutes

----

Liz; Well, that was a hairy one.

East Planner; We aren't really familiar with the traffic. With practice, we could handle that lot.

West Planner; Yes. It would be all right sometimes, but not every day.

Strelsau Approach; Thank God, that looks like weekday peak traffic in August – you don't expect that sort of traffic in November.

Liz; Well done all – see you 1600 – Weekend traffic – Let's see how Hentzau and Zenda get on.

Tuesday 27 1400 Bill; Good Afternoon Team A - Two Sector layout, Weekend Morning 175% Traffic. DUMMKOPF Suspended.

1407 Dr W; ISA – all magenta

1411 Dr W; All ISA Green, East Planner Amber

1413 Dr W; Both Planners Amber

1417 Dr W; Both Planners Red, Both Exec Amber

1420 Dr W; Planners Red, Execs Red, others Green

1424 Dr W; East Planner missed the ISA – make it Red

1431 Dr W; All approach Amber

1437 Dr W; West Exec Amber

1445 Dr W; East Planner Amber Exec Red

1502 Dr W; East Exec Amber

1508 Dr W; West Exec and Planner Amber

1513 Dr W; Strelsau Approach Red

1525 Dr W; Strelsau Approach Amber

1545 Dr W; All winding down Planners Magenta, All others Green

1547 Bill; Thank You Ladies and Gentlemen – Debrief in Conference Room A in five minutes.

Bill; I think that was the toughest exercise I've seen anyone complete. Congratulations.

Strelsau Approach; I lost the picture early on, and never really got it back. I could have used a reliable aid – not bloody DUMMKOPF.

All; It was tough but workable. Not acceptable as a daily level, but occasionally.

Tuesday 27 1500 Wendy; Liz, I have just had the Utopia City Police Department on the blower, of all people. They wanted to know if we knew anything about a person claiming to be "Rupert Hentzauah" from Ruritania Air Traffic Control.

As far as I can make out, he is now in Accident and Emergency at Utopia City Central Hospital, possibly under arrest on suspicion of burglary, but definitely immobilized.

Apparently, they found him about six this morning. It seems a farmer out at Dafurtha Dabetta rang them because of strange noises from one of his pig sheds.

When they got out there, they found a padlocked building – an old farm cottage. The farmer told them he had a sow, a very fierce one, in there with a litter, He had padlocked the door to stop anyone wandering in by chance. They managed to move her into another shed, and found a man suspended below the skylight, by a metal bar that had gone down the back of his trousers. It seems he was obliged to straddle the bar all night, while the sow tried to get at him. He could not reach the skylight, so he had wait until someone heard him.

They had to get the Fire Brigade's platform ladder to get him off. He was semiconscious and unable to stand, so they took him to A & E. Apparently he is not permanently injured, but according to the doctor in charge, he will have eat his meals standing up for a couple of weeks, and "he had better concentrate on thinking pure thoughts for several weeks"..

Liz; Our Rupert has not had a pure thought since puberty.

Wendy; But what was he doing? He's not Eric Gill, after all. Or is he?

Liz; I shudder to think.

Tuesday 27 1600 *Liz; Good Afternoon Team B – Two Sector layout, Weekend Morning 175% Traffic. DUMMKOPF Suspended,*

1608 Dr W; ISA – all magenta

1612 Dr W; All ISA Green, East Planner Amber

1614 Dr W; Both Planners Amber

1617 Dr W; Both Planners Red, Both Exec Amber

1620 Dr W; Planners Red, Execs Red, others Green

1624 Dr W; East Planner missed the ISA – make it Red

1631 Dr W; All approach Amber

1637 Dr W; West Exec Amber

1640 Dr W; Hentzau Approach Red

1643 Dr W; East Planner Amber Exec Red

1645 Dr W; Zenda Approach Red

1702 Dr W; East Exec Amber

1708 Dr W; West Exec and Planner Amber

1730 Dr W; Hentzau Approach Amber

1747 Dr W; Zenda Approach Amber

1750 Dr W; All winding down Planners Magenta, All others Green

1755 Liz; Thank You Ladies and Gentlemen – Debrief in Conference Room A in five minutes

----

Liz; Well, you did it. Well done. How you approach guys feel?

Strelsau Approach; Not as bad as the last one, but I wouldn't care for that every day.

Zenda Approach; No go for routine traffic. I was hanging on by my fingertips.

Hentzau Approach; Me too. Not acceptable for routine operations.

Sector Execs; Really too much traffic.

Liz; Thank you, ladies and gentlemen, see you in the morning. Relax and enjoy your evening. Tomorrow morning will be tough.

Wednesday 28 0900 Bill; Good morning Team B - One Sector layout, Weekday Morning 175% Traffic.- this replaces Monday morning's power cut..

0905 Dr W; ISA running – all Magenta just now

0915 Dr W; All ISA green, Planner Amber.

0920 Dr W All Approach Amber

0922 Bill; Grandad, hold all incoming traffic.

0928 Dr W; Planner is Red, Exec up to Amber

0930 Dr W: Strelsau Approach to Red

0936 Dr W; Hentzau Approach Red

0938 Dr W; Zenda Approach Amber

0940 Bill; Thank You Ladies and Gentlemen. I'm calling a halt, this is clearly not working. – Debrief in Conference Room A in five minutes.

Bill; Well, this is clearly too much. I rather thought so from the start, but we have to prove it several; times to convince the powers that be.

Planner; I'm sorry, Bill, it was just too much too fast.

Bill; Don't be sorry Hans – much better to prove it now than some poor sod having to go through that in real life.

Planner; Especially since that would be me!

(General agreement.)

Wednesday 28 1100 *Liz; Good morning Team A - One Sector layout, Weekday Morning 175% Traffic. This exercise replaces the Monday morning one we lost to the power cut.*

1105 Dr W; ISA running – all Magenta just now

1115 Dr W; All Green, Planner Amber

1120 Dr W All Approach Amber

1122 Liz; Grandad, hold all incoming

1128 Dr W; Planner is Red, Exec up to Amber

1129 Dr W; Exec Red

1130 Dr W: Strelsau Approach to Red

1136 Dr W; Hentzau Approach Amber

1138 Dr W; Zenda Approach Amber

1142 Liz; Thank You Ladies and Gentlemen. Exercise over. Obviously, this is too much all round. Debrief in five minutes.

Liz; I'm sorry to put you through that again, but we have to show the bosses that it won't work even with several tries.

Planner; Should we be able to cope?

Liz; Definitely not. We are trying to show what is acceptable for regular operation, not to ask the impossible, or even the very difficult. You should always have something in hand for emergencies. Not coping is the right thing to do here.

Now go and get some lunch. This afternoon should be easier – very much so.

Wednesday 28 1400 Liz; Good Afternoon Team B - One Sector layout, Weekend Afternoon 100% Traffic. This should be a doddle.

1405 Dr W; ISA running – all Magenta just now

1430 Dr W; All ISA green, Planner Amber

1445 Dr W; Planner are back to Green, Exec up to Amber

1400 Dr W All Green

1430 Dr W: Strelsau Approach to Amber

1436 Dr W; Strelsau Approach Red

1448 Dr W; Strelsau Approach Amber

1550 Liz; Thank You Ladies and Gentlemen – Debrief in Conference Room A in five minutes.

Liz; That was more fun. Any problems?

General Agreement – no problems

Wednesday 28 1500; Jack the Lad;

By the way, Liz, I've been making some enquiries.

Apparently, that waitress Vlada has a way of discouraging unwelcome attentions. She makes an assignation after hours at an old cottage on her uncle's farm – she painted the door white to identify it. Usually, the unwary suitor walks in and falls flat into two feet of pigshit. They normally get the message. She didn't know that her uncle had put the sow in there and padlocked the door, Getting through the skylight must have been Rupert's bright idea.

Apparently, Vlada is the female form of Vlad, and they are now calling her "Vlada the Impaler". That should discourage most nuisances before they start.

Liz; Rupert is going back to Strelsau tomorrow – by train – standing up. I invited him to stay for the farewell party, but he didn't seem keen to appear. The controllers have had a whip-round to buy him a gift, but they seem rather shy about what it is.

Wednesday 28 1600 *Bill; Good Afternoon Team A - One Sector layout, Weekend Afternoon 100% Traffic. This should be an easy one.*

1606 Dr W; ISA running – all Magenta just now

1630 Dr W; All ISA green, Planners Amber

1645 Dr W; Planner back to Green, t Exec to Amber

1700 Dr W All Green

1730 Dr W: Strelsau Approach to Amber

1736 Dr W; Strelsau Approach Red

1748 Dr W; Strelsau Approach Amber

1750 Dr W; Planner Magenta

1755 Bill; Thank You Ladies and Gentlemen – Debrief in Conference Room A in five minutes

----

Bill; Well, that went better – any problems?

No significant problems

Control Room Simulation

Thursday 29 0900 *Liz; Good Morning Team A - Two Sector layout, Weekday Afternoon 100% Traffic. DUMMKOPF Running,*

0905 Dr W; ISA running – all Magenta just now

0930 Dr W; All ISA green, East and West Planners Amber.

0945 Dr W; Planners are back to Green, East Exec up to Amber

1000 Dr W; All Green

1015 Dr W: Strelsau Approach to Amber

1030 Dr W; Strelsau Approach Red

1042 Dr W; Strelsau Approach Amber

1045 Liz; Thank You Ladies and Gentlemen – Debrief in Conference Room A in five minutes.

Strelsau Approach; There is still something very wrong with DUMMKOPF – it is not crashing, but sometimes I wish it would. I think that there is something wrong with its calculation of flight profiles- there seems to be a step change in speed when an aircraft starts to climb or descend.

Thursday 29 1100 *Liz; Good Morning Team B – Two Sector layout, Weekday Afternoon 100% Traffic. DUMMKOPF Running,*

1106 Dr W; ISA – all magenta

1130 Dr W; All ISA Green East Sector Planner Amber

1133 Dr W; Strelsau Approach Amber

1136 Dr W; Strelsau Approach Red

1200 Dr W; All green – going like clockwork – except Strelsau Approach.

1218 Dr W; East Sector Exec Amber

1219 Dr W; West Sector Exec Amber

1221 Dr W; East Sector Exec Red

1230 Dr W; East Sector Exec Amber

1248 Dr W; East and West sectors Exec Green

1251 Dr W; All winding down Planners Magenta

1254 Liz; Thank you Ladies and Gentlemen – Debrief in Conference Room A

Strelsau Approach: Bloody DUMMKOPF. It sits there doing nothing, then suddenly gives a string of warnings, By the time you get to the last one it is nearly too late. Some time it will be.

Liz; Does it give any indication of urgency.

Strelsau Approach; No, it would be a good deal better if it did, although I am not happy even so.

By the way, they have just found the heart-rate monitor – it was in the deep freeze drawer of the fridge – no mouse this time.

Control Room Simulation

Thursday 29 1400 Liz; Good Afternoon Team A - Two Sector layout, Weekend Afternoon 100% Traffic. DUMMKOPF Running,

1405 Dr W; ISA running – all Magenta just now

1410 Dr W; All ISA green, East and West Planners Amber.

1420 Dr W; Planners are back to Green, East Exec up to Amber

1500 Dr W All Green

1503 Dr W; Zenda TMA RED – What's up?

1504 Zenda Approach; What's up Hans – are you all right?

1504 Zenda TMA ; No, dammit, I've broken a tooth and it is hurting like hell.

1504 Liz; Wendy- bandbox your sector with Jack, take Hans to the nurses' room.

Werner, you are not doing anything much – take over Zenda Approach.

Do it now!

1508 Dr W; Zenda Approach Amber

1512 Dr. W; Zenda Approach Green, Zenda TMA Amber

1515 Dr W; Zenda TMA Green – Well done, Werner.

1530 Dr W: Strelsau Approach to Amber

1536 Dr W; Strelsau Approach Red

1548 Dr W; Strelsau Approach Amber

1550 Liz; Thank You Ladies and Gentlemen – Debrief in Conference Room A in five minutes.

-- -- ---

Liz; Well, Werner, that was a good piece of fielding.

Wendy is taking Hans to the dental hospital.

Strelsau Approach; Sorry to be so negative but DUMMKOPF really is getting in the way.

Thursday 29 1600 Bill; Good Afternoon Team B – Two Sector layout, Weekend Afternoon 100% Traffic. DUMMKOPF Running,

1606 Dr W; ISA running – all Magenta just now

1630 Dr W; All ISA green, East and West Planners Amber.

1645 Dr W; Planners are back to Green, East Exec up to Amber

1700 Dr W All Green

1730 Dr W: Strelsau Approach to Amber

1736 Dr W; Strelsau Approach Red

1748 Dr W; Strelsau Approach Amber

1755 Bill; Thank You Ladies and Gentlemen – Debrief in Conference Room A in five minutes

-- -- ---

Bill; Well, that was a smooth one – good work.

Strelsau Approach; I have to say again, I do not trust DUMMKOPF. I cannot put my finger on the problem, but there is definitely something wrong.

Friday 30 0900 Bill; Good morning Team B - One Sector layout, Weekday Morning 100% Traffic.- this replaces Wednesday morning, which re-ran Monday's power cut.

0905 Dr W; ISA running – all Magenta just now

0925 Dr W; All ISA green, East and West Planners Amber

0930 Dr W; Planners are back to Green, East Exec up to Amber

0937 Dr W; All Green

1015 Dr W: Strelsau Approach to Amber

1030 Dr W; Strelsau Approach Green – everyone Green

1045 Bill; Thank You Ladies and Gentlemen – Debrief in Conference Room A in five minutes.

Strelsau Approach; I think the problem with DUMMKOPF in its present state is that you have to keep three pictures. The usual one for approach, another one for what DUMMKOPF is suggesting, and a third for what to do if DUMMKOPF goes belly-up.

Rather than saving effort, it trebles it.

Bill; Final general de-brief Conference room B (next door)1300 hours, with Team A – bring all your worries.

Friday 30 1100 *Liz; Good morning Team A - One Sector layout, Weekday Morning 100% Traffic.- this replaces Wednesday morning, which we used to replace Monday's exercise we lost to the power cut..*

1105 Dr W; ISA running – all Magenta just now

1130 Dr W; All ISA green, East and West Planners Amber

1145 Dr W; Planners are back to Green, East Exec up to Amber

1200 Dr W All Green

1215 Dr W: Strelsau Approach to Amber

1230 Dr W; Strelsau Approach Green

1245 Liz; Thank You Ladies and Gentlemen – Debrief in Conference Room A in five minutes.

-- -- ---

No significant problems

Liz; You should know that Hans's tooth problem has been sorted out. It involved a good deal of digging. He was pretty groggy when Wendy took him back to his hotel. Fortunately, his wife was there, and she said she was taking him home this morning,

He will miss the wind-up party, but that is probably just as well.

Liz; Final General De-brief in Conference Room B (next door) at 1300 hours. If there is anything more you want to talk about, now is the time. .

Day of Week/ Session	Monday 12/10 Training	Tuesday 13/10 Training	Wednesday 14/10 Measured	Thursday 15/10 Measured	Friday 16/10 Stand-by
0900-1100	Arrival	A2WkAM- 80%	~~B1WkAM- X 100%~~ Staff Prob	A2WkPM- O 100%	B1WkAM- X 100%
1100-1300	Introduction	B2WkPM- 80%	~~A1WkAM- X 100%~~ DUMMKOPF	B2WkPM- O 100%	A1WkAM- X 100%
1400-1600	A1WkAM- X 50%	A2WeAM- 80%	B1WePM- O 100%	A2WeAM- O 100%	Free
1600-1800	B1WkPM- X 50%	B2WePM- 80%	A1WePM- O 100%	~~B2WeAM- O 100%~~ Data Lost	Free
Day of Week/ Session	Monday 19/10 Measured	Tuesday 20/10 Measured	Wednesday 21/10 Measured	Thursday 22/10 Measured	Friday 23/10 Stand-by
0900-1100	B1WkAM- 125%	~~A2WkPM- X 125%~~ DUMMKOPF	B1WkAM- 150%	A2WkPM- X 150%	A2WkPM- X 125%
1100-1300	A1WkAM- 125%	B2WkPM- X 125%	A1WkAM- 150%	B2WkPM- X 150%	B2WeAM- 100%
1400-1600	B1WePM- 125%	A2WeAM- X 125%	B1WePM- 150%	A2WeAM- X 150%	Free
1600-1800	A1WePM- 125%	B2WeAM- X 125%	A1WePM- 150%	B2WeAM- X 150%	Free
Day of Week/ Session	Monday 26/10 Measured	Tuesday 27/10 Measured	Wednesday 28/10 Measured	Thursday 29/10 Measured	Friday 30/10 Stand-by
0900-1100	~~B1WkAM- 175%~~ Power Cut	A2WkPM- O/L 175%	B1WkAM- O/L 175%	A2WkPM- X 100%	B1WkAM- 100%
1100-1300	~~A1WkAM- 175%~~ Power Cut	B2WkPM- 175%	A1WkAM- O/L 175%	B2WkPM- X 100%	A1WkAM- 100%
1400-1600	B1WePM- O/L 175%	A2WeAM- 175%	B1WePM- 100%	A2WeAM- X 100%	Farewell
1600-1800	A1WePM- O/L 175%	B2WeAM- 175%	A1WePM- 100%	B2WeAM- X 100%	Departure

X = DUMMKOPF Run OK   O = DUMMKOPF Not Run   Blank = not planned
~~X~~ = DUMMKOPF Malfunction   O/L = Overload – exercise halted

Figure 10 - Final Experimental Plan

Final De-Briefing

Liz; Well, I'll begin with the big problem – DUMMKOPF (Groans). Now be fair, Igor here has worked day and night to get it working, and his efforts deserve respect. (Sorry, Igor).

Clearly, DUMMKOPF is not ready for actual deployment. However, if the technical problems can be ironed out, would it be useful?

(Various controllers, condensed). As it is you have to work on the approach job, on understanding what DUMMKOPF is doing, and on deciding what to do if it keels over.

IF: -

- *DUMMKOPF was completely separate from the main operating system.*
- *DUMMKOPF was fully tested.*
- *Controllers were involved in a systematic evaluation of DUMMKOPF.*
- *DUMMKOPF could be asked to explain its decisions.*

A Real-Time Simulation is not really the right place to test this sort of software. You should rig up a simple fast-time simulator MicroSAINT (David, 1997) would probably do, if you could get the source code. Then you should systematically explore the working of DUMMKOPF, trying out extreme values – high landing speeds or helicopters – in sub-zero or tropical atmosphere conditions, gale-force crosswinds and so on. You will be looking for "one in a million" occurrences, and no-one can afford to run several million hours of Real-Time simulation, which you would need to turn them up. Any reliable checking system will look at things like checking that divisors can never be zero, that functions are never called with impossible parameters, and things like that. If you are feeling sneaky enough, Igor, you can incorporate code that is never used, but, for example takes the square root of a negative number, or the reciprocal of zero. See that the checking software catches it. If it doesn't, ask for your money back. If it does, see what it does about it. Simply stopping is not enough.

If you can solve this lot, then you would have a very useful tool. The fundamental principle is unorthodox, but the logic is impossible to deny.

If you do get it working, produce a re-play version that can explain why it makes the decisions it does. That would be really useful. It might even teach Hans something.

Liz; Well that wraps that up. Now about the two-sector split.

(Again, controllers condensed) The two sector split cuts down the peak workload for Planner and Executive, but it makes a good deal of work between the two sectors. (Some controllers doubt if it is worth it. Others feel that something has to be done, and this seems the only reliable possibility. A minority feel that better, more streamlined, methods and updated display software would produce the same effects.)

Liz: how much capacity could you gain with the two sector layout on week days – and weekends.

First off, how much capacity have you got with the present single sector layout?

(Controllers' opinions summarised)

We can handle 125% easily. We can handle 150% Weekday traffic, but the 150% Weekend traffic is too much. We would have to apply some sort of flow control, although this was supposed to be 95% traffic, so that would mean about 20 days per year. 175% is definitely too much. Say 150% of Weekday traffic and 140% of the current Weekend.

And how much with the two sector layout.?

(Controllers' opinions summarised)

Again 125% is easy, and 150% is just about workable, with more practice. 175% we could just cope with for weekday traffic, but not for weekends. If you said somewhere about 170% for Weekdays and 160% at Weekends, you would not be far wrong.

Any other overall comments?

(About half the controllers)

Our present equipment is out of date. It is high time we got up-to-date displays and something better than strips. We need a conflict detection system, one that takes account of aircraft intentions, not a simple extrapolation – w tried that and it gave too many false alarms. .

And we need to look at some of the more common emergencies as well.

Liz; Finally, I have here a large ashtray with a candle in the middle. Please burn your identity number card. If you want to keep it as a souvenir you can, although I have here a little book about the Experimental Centre for each of you – with a personal letter of thanks from the director. Can someone make sure Hans gets his copy, with our best wishes? - Thanks, Hildegard. .

Farewell Party

Friday 30 1330 Conference room B

Liz; Ladies and Gentlemen, may I have your attention please. First, on behalf of the EREWHON EXPERIMENTAL: CENTRE, and of its staff, I thank you for your contributions to the successful completion of the RURITANIA I Simulation. I can truly say that without your cooperation, this simulation could not have taken place. I hope you have enjoyed it as much as we have. (laughter) You already know that the conclusions will be taken very seriously where it matters, and implementation of its recommendations should begin very soon – by ATC standards. (Groans)

I must convey the apologies for absence of Hans Brinker – his tooth trouble has been dealt with, but we felt he should go home as soon as possible – and of Rupert Hentzau (Laughter) – he intended to drop in, but was unfortunately held up (Renewed Laughter).

After discreet enquiries, we thought it better not to offer you our version of KartoffelSuppe – (Omnes : Kartoffelsuppe, Kartoffelsuppe – Immer Verdammte Kartoffelsuppe) - but to provide a buffet of our best Utopian dishes. This, I am assured, will be an unforgettable experience (Cheers!).

So, without more ado – dig in!

Rabbit, Rabbit, Rabbit, clink, clink, clink, clink, gurgle, gurgle, gurgle, gurgle, gurgle

(Hic!). (da capo, ad nauseam)

Strelsau Approach (A) ; Splendid party Liz – We were going to have a presentatation to Rupert in recognition of his adventures – A nice framed picture and an illuminated assress – addedds – dammit address! We are going to present it now, and forward it ananym – anomon – ammon – dammit without saying who it's from.

Liz; I'm sure he will take it in the spirit it is meant, but just in case, I'm going to Nephelocccygia this weekend. I'll post it from there, if you like.

> Randy dandy Rupert Hentzau
> On lechery was intent, so
> He fell through a skylight
> And ended the night right
> On top of an angry immense sow!

Go for the low-hanging fruit, Mama

(Hmm. Not exactly deathless verse, nor immortal art, but the meaning is all too clear. There will be the devil to pay if this gets out. I had better make a few photocopies – you never know.)

You know, Liz, we've been looking around while we were here, and your people have been showing us what is going on elsewhere. We had no idea. Strips have been around since the year dot. It really is time we moved on. A lot of these new displays are just patches to replace strips with data displays. One of them had some old papers by someone called Hugh David. Apparently, they were published by an obscure Institute, and sank without trace, although I think he put together a book, which may have been published.(David, 2018) Have you ever seen a copy? He must be the only octogenarian teenage radical around, but some of his ideas make sense. This damned radio link is archaic – surely we can do better and safer. He has some radical suggestions on display and control, and on the whole caboodle. In fact he calls the whole business the Air Traffic Kludge.

Liz; What is a Kludge?

Strelsau Approach (B); Apparently a system thrown together with no overall plan, accumulating bits from all over the place. It works, but nobody knows why or how it will finally collapse.

Liz; Sounds like Air Traffic Control – present company excepted of course.

Strelsau Approach et al; Not by us!

(General song;)

"Yr wyf wedi blino ar. Mae fy nhraed yn brifo. Mae gennyf ddannodd.
Rydw i wedi blino. Yr wyf yn llwglyd, ac yr wyf am fynd adref.
Pam ddiawl oedd erioed i gymryd rhan mewn prosiect gwaedlyd hwn?
Pwy sy'n y uffern yn gofalu beth sy'n mynd ymlaen yn efelychiadau amser real?
Nid yw'r bodoli. Mae neb yn hoff iawn imi.
 Gollfarnu hyn i gyd.
Gollfarnu hyn i gyd."

Liz; What the hell is that?

Flavia Elphburg; That is our national anthem in the traditional Old Ruritanian.

Liz; Very noble, very impressive!

(Some time later)

Liz; Well, Bill there they go. A good lot, on the whole. Wendy, did you collect the parking permits? Oh, damn! You can ask Admin to write to ask Hentzau to collect them and send them back. I'm sure the Controllers will be glad to see him.

I'm off for a week in the sun – then I'll finish the report. So you lot can start on the Kennaquaire III simulation –Charlie Stewart and Flora Macdonald will be running it. Best of luck. At least you should get a decent drink on that one.

Chapter 13 – Analysis

Ideally, the results of a simulation exercise or a real-life trial should be subjected to statistical analysis. It may come as a shock to academic readers to learn that this is often not the case. For many years, it was customary for Real-Time Control Room simulation leaders to make recommendations based on their impressions only, taking no account of the statistical significance of the results observed. Even where statistical analyses were carried out, they were generally used, in the words of Andrew Lang, "as a drunk uses a lamp-post - for support, rather than illumination".

There was something to be said for this point of view, since the quality of the statistical analysis was not high, and many simple confounding factors were neglected. It was common practice in large Real-Time Control room simulations to begin by simulating the existing system, then to proceed to the proposed options. (The confounding of learning effects is obvious.) Where one possible option was obviously unworkable, it was usually abandoned and an improvised alternative inserted in its place. The 'balanced experimental design' beloved of statisticians and necessary for correct analysis of variance went by the board. Frequently, all operators were regarded as equally competent, and carefully balanced seating plans would unravel as soon as the program of exercises started.

Fortunately, the observed differences were usually so great that no real question arose over the practical recommendations. Indeed, by the time the formal simulation report had been written, edited, printed, bound and distributed, its recommendations had been put into practice. Such reports were usually of restricted circulation, and rarely subject to any scientific review. There is an uneasy suspicion that most copies of formal reports are received by those on the distribution list, weighed in hand and then shelved, without ever being actually read. Others may face, so to speak, a more fundamental fate.

Dopey ; So, thank you very much, Liz. Enjoy your holiday.

Now, Let's see.... Wendy, get me the final experimental plan, and the two Massie Grids. Then you and Petra can get on with data checking .

Statistical Analysis

In the contemporary world and (one hopes) in the future, more scientific methodology may be usefully applied. Computer-based statistical analysis packages are available, and may be almost as useful as their promoters suggest. Some general guidelines may be useful to the non-statistician. Professionals should not need to be reminded of them, but...

Statistical analyses are divided into 'descriptive' and 'analytic' types.

Descriptive statistics present data in agreed forms to assist the user in seeing what is happening or has happened in the course of a data gathering exercise. Ideally, they should show what is important, while suppressing what is not.

Analytic statistics attempt to determine whether observed differences are sufficiently large to represent real underlying differences, rather than mere accident. To do this they usually make some assumptions about the data. Most commonly, using 'parametric' statistics, they assume that data, apart from the systematic differences they are testing, are 'normally distributed'. The 'normal distribution' is the so-called 'bell shaped curve', which can be described as the limit of the sum of a large number of small random differences. Engineers use this distribution because they believe it is theoretically justified, while theoreticians use it because they believe it is practically justified. Both are wrong.

Analytic statistics are highly regarded in the scientific community, and many valuable lives have been spent in elaborating them. They are usually redundant in many real-life situations, and can in fact be dangerously misleading.

If you do not yourself know about these fascinating/boring aspects of analysis, and have neither the time nor the inclination to spend several years finding out about them, make sure you are employing someone who has. As an empirical check, ask them about the effects of confounding, non-normality and heteroscedasticity. If they cannot answer you, you - and they - are in trouble.

Data Checking

First, checks should be made that recording is taking place. It is surprising how often a data file turns out not to have been recorded, or to have been overwritten, lost or accidentally destroyed. It may be possible to re-run an exercise to replace lost data, if it is spotted in time.

Friday 23 1100 Liz; Good Morning Team B - Two Sector layout, Weekend Morning 100% Traffic This is a repeat of last Thursday afternoon – some silly sod overwrote the records so we have to do it again. The traffic is still 100%, but it will be different from the previous time. This should be an easy one.

If it is not spotted in time, you will have to find ways of adapting your analysis. This is where a competent statistician should be able to help.

A very important preliminary step in the analysis of data is the identification of 'outliers' – data points which are widely different from the rest. The simplest method of identification is to plot the data and look for any points that are well away from the rest. (One simple cause of errors that occurs more often than might be expected is the 'missing datum' recorded, for example, by an automatic eye-tracker when the subject blinks. This may be recorded as a '0, 0' pair of coordinates, and may cause strange distortions of the 'point-of-gaze' data.) If something dreadful happens during a simulation, the controller may forget, or have no time, to respond to the ISA prompt. This will usually result in a maximum response time and a zero score.

Most statistical packages do not attempt to detect or warn the user of possible anomalies in the data. As a rule, all data should be plotted, preferably 'on-line' so that corrective action may be taken. If this is not possible, numerical data should be listed that are more than five standard deviations from the overall mean, or in classes containing less than one per cent of the data. For ordinal (ranking) data one should look at any classes containing only one reading, since this often a slip of the recording. It is not wise to reject such data automatically, since the causes and distribution of anomalies may provide important cues to oddities of system behaviour. Rather off our current topic, the existence of the 'ozone hole' in the upper atmosphere was recorded long before it was identified, but the recordings were automatically discarded by data checking routines.

Unfortunately, it is not always possible to identify small errors, which may hide in the general variation present. Fortunately, in this context, small errors do not matter much. Equally important in the real world is the unambiguous identification of files of records, of which there may be dozens or hundreds in a large-scale simulation. All files should contain at least the date and time of origin, and any exercise identification code used.

Petra; Why do we get the boring jobs?

Dopey; Because you do them so well, dear.

Wendy; Perhaps we had belter start making mistakes on purpose.

Dopey; A boring job is better than no job – and a bad reference.

Petra, check the files. Wendy, start looking for anomalies, you can use the auitochecker as Petra clears the files.

Off you go.

- - -

Petra: Including the overloads, but leaving out training, we have eight exercises in week 1, and sixteen each in weeks 2 and 3. That gives 40 exercises, of which five in week 3 were overloaded.

For half the exercises we have eight working positions, and for the other half we have ten. We should have a total of 360 NASA-TLX and end-of-run questionnaires.

For both teams, we should have ten End-of-Simulation questionnaires and ten Operator characteristics questionnaires.

Wendy: We are one short on the End-of-Simulations – that must be the bloke who broke a tooth . Yes, Thursday 29th 0900. And we are five short on the Operator characteristics. (It wasn't compulsory, anyway.)

We have complete event records for the 35 completed exercises, and stubs for the five overloaded ones.

Petra; We have heart rate records for the first week, and the first two exercises of week two. After that, it went missing ... Half of those were training, and the rest were baseline exercises.

Wendy; Check with Dopey, but we had better scrap heart rate.

The hard working and dedicated assistants now spend some considerable time converting this data to columns in a set of Microsoft EXCEL spreadsheets.) They then make plots of each column and look for exceptional values. Most exceptional values turn out to be input errors, but some require recourse to the running diary, or to anyone (Bill Darcy, for example) that they can get hold of.

Simple Data Analysis

It is usually good practice to analyse each type of data individually, using an appropriate test for the level of data employed. (Yes/No responses cannot be analysed in the same way as readings of heart rate, for example.) Although the nature and details of statistical tests are beyond the scope of this text, most tests report the probability that the observed differences, or more extreme differences, are likely to occur if there is in fact no real difference. If this probability is below an agreed level, usually 5% or 1%, then it is accepted that a difference exists. If you have a large number of independent data streams – for example one thousand, then you will find about fifty differences that are reported as significant at the 5% level and ten at the one per cent level if there is no real effect present. If the thousand data streams are not really independent, the number may be more or less. A lower probability for accepting a difference – say 0.1% is probably the best response. An alternative may be to split the data into two groups and see if the significant differences occur in both groups. This is a difficult topic, and you should consult a competent statistician, (If you can find one!).

It is often useful to present results in a form that shows how they relate to normal, acceptable or limiting levels of what is being measured. For example, in traditional Air Traffic Control, it would be considered normal for a Radar Controller to spend 30 % of his time talking on the frequency, 40% being the acceptable level and 50% the limit. These limits are more or less arbitrary, and might be expressed as 32%, 43% and 49% to impress the reader with their precision.

It is tempting to rely only on mean values, but the distribution of values can be equally important, - the average citizen of the United States of America has one breast and one testicle. Equally, the maximum or 95%ile value may be more informative in judging the acceptability of performance. (The maximum, incidentally, is unusual in that the more data you have, the less reliable it becomes.)

Analysis of Variance

Although Analysis of Variance was defined by R. A. Fisher in 1925, I was told it was 'too advanced' for Air Traffic Control in 1970.

Analysis of variance is the workhorse of statistical analysis. It is very widely used, and sometimes understood. It requires a systematic experimental plan, and, if there are holes on the plan, (missing exercises, unbalanced numbers in different cells of the design, or just plain mistakes) it can be extremely misleading. Some of the potential errors can be detected automatically, although many 'statistical packages' will not warn you of them. There are ways of getting round most of these, but I should not be taking the bread from the mouths of my fellow statisticians.

This technique applies to values on a continuous numerical scale. Taking, for example, the percentage of time spent talking to aircraft in each sector in each run of a simulation, it is possible to estimate the amount of variation contributed by the sector, the organisation and the remaining unexplained variation. A statistical test can be used to estimate the chance of observing as much or more variation assigned to sectors and organisations. However, there are many pitfalls in this process. For example, the unassigned variation (the Residual) is assumed to be a measure of unexplained 'random variation', and to be 'normally distributed.' (don't ask). Neither is necessarily true, and there are ways of dealing with these aberrations. Another common fault is to have uneven numbers of values, where some runs have failed or not been recorded. Here again, there are remedies, but many packages do not warn the unsuspecting user.

If you have what is called a 'balanced design' you can look at the overall variation (variance) of the data and extract systematic effects of, for example, traffic density, simulated time of day, simulated weekday or weekend, and so on. We call these 'factors'. You can also extract 'interactions'. These can show, for example, that one particular combination – such as 'Weekend afternoons' is worse than you might expect from the overall 'Day of Week' and 'Time of day' effects. More probably, the traffic loading will affect the 'Day of week' or 'Time of day' effect.

We end up with a list of Factors – what we varied - and a list of variates – what we measured.

Analysis of Variance shows how we can distribute the variation in each variate by factors and their interactions.

Dopey: I think I need a coding for the things we varied.

Code	Characteristic	Value	Value
Team	Team	A = First Team	B = Second Team
Sect	No of Sectors	1 = One Sector	2 = Two Sector
ToD	Time of Day	1 = AM = Morning	2 = PM = Afternoon
DoW	Time in Week	1 = Wk = Weekday	2 = We = Weekend
DK	DUMMKOPF	1 = X = Used	2 = (blank) Not Used
Traf	% Current traffic	50/80 Training	100/125/150/175 Measured
TwR	Time within Run	Value in 3 min intervals from Start 1 to 30	

Figure 11 - Factor Coding

Dopey; Where did we have overloads? Wednesday 20th All one sector 150% , Monday 26th All one sector 175%, including the two that were re-run on Wednesday 28th.- and one two sector 175% on Tuesday 27 morning. (Check these Petra).

So – we can discard the Monday 12th and Tuesday 13th training exercises. Wednesday14th, Thursday 15th 100% (Including the re-plays. OK

19th Monday and 20th Tuesday 125% OK

21 Wednesday 150% One Sector Not OK and 22 Thursday 150% 2 sector OK

26 Monday 175% One Sector all four Not OK, 27 Tuesday 175% Two Sector 3 out of 4 OK

28 Wednesday and 29 Thursday 100% OK Usual Pig's breakfast.

So we have 100%, 125%, 150%(less1), 175% (less 5) better leave out the 175% for analysis, but sort out data anyway, for the ones we have. There is nothing we can get from the overload records – no, hold on, we do have the NASA-TLX and comments.

And separately we have 100% at the start and at the end – that will do as a check on learning during the simulation, compared with the first set of 100%

Now what about DUMMKOPF. The one or two sector factor does not matter, and we are only interested in Strelsau TMA. We should try to get a balanced set of trials. We will go into that later – lets sort out the sector effects first..

Before we plunge into analysis, we need to consider what we are interested in comparing.

Working Position. *We have Planner and Executive for Overall, East Sector, and West sector. However, we have Overall only for One Sector, but East and West only for Two Sector. So what do we do?*

We could compare the average workloads by using Overall, East and West in place of Sect. We can subdivide the three-way comparison which has two degrees of freedom into Overall vs the rest with one degree of freedom, and East Versus West. This is an orthogonal partition, and if you do not know what that means, you are not the only one. Practically, this means running the Anovar (Analysis of Variance) with all three, then with only East and West. Subtracting the sums of squares and degrees of freedom of the second analysis from the first gives the differences of overall from the average of East and west.

Alternatively, we could go through all the Excel tables averaging the East and West values, which could be compared with the Overall values to produce the same result. That could also be used to compare the totals for East and West with the Overall values. Things like total aircraft entering and leaving would make sense, but the total ISA would mean nothing at all. Have to go through the Massie Grid to check which we need.

So that gives an anovar with an extra factor

Code	Characteristic	Value	Value
Pos	Type of Position	1 = Planner	2 = Executive

Figure 12 - Sector Position Factor

Missing Values. *If we leave out the 175% cases, we have 100%, 125% and 150% levels of Traffic for each Team, Time of day and day of week.*

We can now make up our analysis of variance table – we will put in the values as we do each variate, but the degrees of freedom will be the same.

Control Room Simulation

Source of Variation	Degrees of Freedom
Team	1
Sectors	1
Position	1
Day of Week	1
Time of Day	1
Traffic Level	2
Time within Run	24
Residual	2368
TOTAL	2399

Figure 13 – Sectorisation Anovar (Main Effects)

This gives us the main effects.

Before we go on better look at these.

Team – *is not important for the report, since the two teams were simply chosen as representative, with nothing special about them. We keep it in the analysis simply to check that there are not significant differences. (If we leave it out, and there are significant differences, it would make the residual bigger and the significance of the things we are interested in would be reduced.)*

Sect – *is what this is all about. The difference between one sector and two-sector is what we are looking for.*

Posn – *the differences between the effect on Planner and Executive are going to be important, since their workloads peak at different times.*

DoW – *the Day of the week (Week-day or Weekend) is basically what we are being asked about. The original Question asked was about the capacity at weekdays and weekends separately. In fact this is not a factor for analysis but a data separator. We will; be doing our analyses for WeekEnd and WeekDay separately.*

ToD – *the time of Day (Morning or Afternoon) is there as a precaution. It may be that one or the other is significantly more difficult.*

Traf *– the traffic level (100, 125 or 150%) is essential. The interactions of this effect with other effects give a clue when a Sectorisation is reaching its limit. (We have two replications of 100%, one at the start and one at the end. We have to choose one replication to keep the design balanced. Although we could compare the two 100% levels separately to see if there is a significant learning going on, we will choose the second iteration – at the end of the simulation, for comparison.)*

TwR *We have 90 minute runs, giving us 30 readings at three-minute intervals within the run, but these start with an empty system. We should discard the first 15 minutes, to get a clearer view of the 'steady state'. We do not have a 'run-down' at the end of each run, since they were stopped by the simulation leader. So we use values from 6 to 30 (25 per run) for all the 3-minute values (including ISA)*

This means that the differences between successive three-minute intervals are used as a measure of the overall residual variation, which ANOVAR assumes is random. If you look at the ISA records, for example Figure 14, you see that ZA (Zenda Approach) starts at 2 (Cyan, meaning 'Low' from reading 6 to reading 10 but levels off at 3, the numerical equivalent of Green meaning 'fair workload' for the remaining readings. Clearly these are not independent values, and will over-estimate the significance of differences. (If you do not see why, ask a statistician.) Moreover, some of the variates are not measured at three-minute intervals, so they only have a single value per exercise. Figure 14 shows for which of the variates in the main Massie Grid we have 3 minute or run values. We have 14 variates at 3 minute intervals (one of which we did not get – HR) and 16 for the whole exercise (of which one (TLX) we do not need, one (PHR) we did not get and one (NMA) which has nothing to do with the Sectorisation – it has a lot to do with DUMMKOPF, but that is another story.) 27 variates in all!

Damn it!

The simplest thing to do is to take the mean values for the remaining 3 minute interval variates and come back to look at the inside run effects later – if we have time

So we now have the main effects as shown below. (Figure 15).

Measure	Time scale
EAC	3 min
LAC	3 min
PAC	3 min
NFREQ	3 min
TFREQ	3 min
PFREQ	Run
NINTER	3 min
TINTER	3 min
PINTER	Run
NHAND	3 min
THAND	3 min
PHAND	Run
NORD	3 min
TORD	3 min
PORD	Run
XORD	Run
VISA	3 min
TISA	3 min
NC5	Run
NC2	Run
NMA	Run
TLX	Run
TLXM	Run
TLXP	Run
TLXT	Run
TLXQ	Run
TLXE	Run
TLXV	Run
HR	3 min
PHR	Run

Figure 14 – Duration Of Measure

Factor	Degrees of Freedom
Team	1
Sect	1
Posn	1
ToD	1
Traf	2
Residual	41
TOTAL	47

Figure 15 - Run Means Anovar Main Effects

We have eight exercises at each of three levels, with two different positons, giving us 48 readings. ANOVAR gives differences around the overall mean, so we have 47 degrees of freedom.

but we are going to need some of the interactions.

The total number of two-way interactions is 5! / (2! x 3!) = 10

The total number of three way interactions is 5! / (3! x 2!) = 10

The total number of four way interactions is 5! / (4! X1!) = 5

(The general formula where n is the number of Main effects and k is the number of effects in the interaction is n! / (k! x (n-k)!) where ! is the factorial sign - so 3! = 3 x 2 x 1)

We are not interested in a lot of these interactions. To start with, the Team effect (Which Team of controllers?) is not important, so we can cut out all interactions involving it.

Second order interactions

Sector by position is important – is the Sectorisation more important for the Planner or Executive.? Keep that (1 degree of freedom)

Sector by Time of Day – also important. Keep. (1 degree of freedom)

Sector by Traffic load – Keep (2 degrees of freedom)

Position by time of day - ? Keep (1 degree of freedom)

Position by Traffic load – Keep (2 degrees of freedom)

Time of Day by Traffic load – Keep -(differences are more likely to show up under heavy load. (2 degrees of freedom)

Third order interactions *(ignoring Te again)*

Sect x Posn x Td (working position by Time of day) ?

Sect x Posn x Tr (working position by Traffic Load) Keep

Sect x Td x Tr (Sectorisation by time of day by Traffic Load) ?

Posn x Td x Tr(Position by time of day by traffic load) ?

Fourth order *interactions (still ignoring Te)*

What the possible use of Sect x Posn x ToD x Traf might be baffles me.

Some ANOVAR programs will combine the non-significant interactions with the residual, reducing the volume of output considerably. Otherwise it can be done by hand. So now we have the following full ANOVAR (Figure ??)

(I have a feeling that some of these interactions are confounded – better check when we have the means)

So now we can run a first try of the anovar program. This gives us 27 x 2 analyses. Each with a total of 15 sources of variation. 810 different tables of means…

Liz will love that. We had better highlight any that are significant – and fill in the Massie grid accordingly – that is what it is for after all, and we can add a brief explanation column

Source of Variation	Degrees of Freedom
Team	1
Sect	1
Posn	1
ToD	1
Traf	2
Sect x Posn	1
Sect x ToD	1
Sect x Traf	2
Posn x ToD	1
Posn x Traf	2
ToD x Traf	2
Sect x Posn x ToD	1
Sect x Posn x Traf	2
Swct x ToD x Traf	2
Posn x ToD x Traf	2
Residual	25
TOTAL	47

Figure 16 - Full Anovar - Sectors

Measure	1 vs 2 Sectors Weekday	1 vs 2 Sectors Weekend	Sum/ Mean/ Peak?	Comment
Entry AC	++	++	Sum	Expected
Leave AC	++	++	Sum	Expected
Present AC	=	=	Sum	Expected
No. FREQ	+	+	Sum	Expected
Tot FREQ	+	+	Sum	Expected
Pct. FREQ	-	-	Peak	Expected
No. INTER	+	+	Sum	Expected
Tot INTER	+	+	Mean	Expected
Pct. INTER	=	=	Peak	Poor Data
No. HAND	+	+	Sum	Expected
Tot HAND	=	=	Mean	Expected
Pct. HAND	=	=	Peak	Poor Data
No. ORD	+	+	Sum	Expected
Tot ORD	+	+	Mean	Expected
Pct. ORD	+	+	Peak	Expected
XORD	=	=	Sum	Poor Data
Val ISA	--	--	Mean	Expected
Time ISA	=	=	Mean	Expected
NearC5	=	=	Sum	Poor Data
NearC2	=	=	Sum	Poor Data
TLXMent	--	--	Mean	Expected
TLXPhys	=	=	Mean	Expected
TLXTemp	--	--	Mean	Expected
TLXPerf	=	=	Mean	?
TLXEffort	--	--	Mean	Expected
TLXFrust	--	--	Mean	Expected

-- = 1 higher (<0.1%)   - = 1 higher (<1%)   = = No Sig difference

+ = 2 Higher (<1%)   ++ = 2 Higher (<0.1%)

Figure 17 Sectorisation Effect

Expected = we have a significant difference (or none) in the expected direction

Poor Data = Not enough data to make a statistically valid judgement.

? = unexpected result.(Probably muddled meaning?)

Now DUMMKOPF – What? Damn! I have to go to the Kennaquhair planning meeting – back soon – I hope!

Later (much Later) Dopey; Where were we? We have the TMA and Approach controllers for Hentzau, Strelsau and Zenda for all exercises. We can split them into two factors, and add them to the table of factors.

Code	Characteristic	Value	Value	Value
Posn	Working Position	1 = TMA	2 = Approach	
Locn	Location	1= Hentzau	2 = Strelsau	3 = Zenda
Team	Team	A = First Team	B = Second Team	
Sect	No of Sectors	1 = One Sector	2 = Two Sector	
ToD	Time of Day	1 = AM = Morning	2 = PM = Afternoon	
DoW	Time in Week	1 = Wk = Weekday	2 = We = Weekend	
DK	DUMMKOPF	1 = X = Used	2 = (blank) Not Used	
Traf	% Current traffic	50/80 Training	100/125/150/175 Measured	
TwR	Time within Run	Value in 3 min intervals from Start 1 to 30		

Figure 18 - TMA Factors

Come to think of it, we can drop the Hentzau and Strelsau data first time, since we are strictly looking for the DUMMKOPF effects. It may have a slight overflow effect on the others, but it is Strelsau that matters. In fact this is really a data separator – we are not really interested in the average TMA behaviour.

Looking at these again

Posn *is obviously important. The Approach controller is the one who is supposed to be using DUMMKOPF.*

Locn *is a data separator this time. We have no real reason to be interested in Hentzau or Zenda, although we should repeat the analysis for them later.*

Control Room Simulation

Team, Sect, Tod and Dow *are all essentially control variables*

Traf *is obviously important, Does DUMMKOPF give better results with heavier traffic – or worse?*

DK *of course is what it is all about.*

Traf *will affect the ratings – especially NASA-TLX and ISA, but will it affect them more in heavy traffic?*

TwR *- the time within the run will certainly affect most things, where we have three minute intervals, but we had better work with the means of runs for a start. – so TwR can go out of the analysis.*

However, we have a problem – DUMMKOP kept going wrong so we do not have a full set of runs with it – let's look at the actual exercise record – Figure 10

What a bloody mess!

Week 1 – ignore the 50% and 80% - they were training. That leaves One Sector Weekend Morning at 100% for both teams on Friday with DUMMKOPF – we could compare with Wednesday 14 afternoon (One sector Weekend Afternoon), Thursday 15 morning, (Two sector Weekday Afternoon), or Thursday 15 afternoon (Two sector Weekend Morning) – which is most like? I think Thursday Afternoon – One or two sectors should not affect Approach control different traffic probably would.

Week 2 - .Not so bad. We can use all eight 125% exercises and all eight 150% exercises. We have 1 sector and 2 sector confounded with using DUMMKOPF, but what the hell.

Week 3 175% is a total pig's breakfast. The 100% second set is all OK, so we can use these instead of the first week.

So we have three complete repetitions (with 100% 125% and 150% traffic)

DUMMKOPF	No DUMMKOPF
A2WkPM	B1WkAM
B2WkPM	A1WkAM
A2WeAM	B1WePM
B2WeAM	A1WePM

So Sector is confounded with DUMMKOPF – shouldn't matter

DUMMKOPF has WeAM and WkPM, while NO DUMMKOPF has WePM and WkAM – could be a confounded nuisance.

So what have we got left? Let's try an Anovar table?

Source of Variation	Degrees of Freedom
DUMMKOPF (=DK)	1
Posn	1
ToD	1
Dow	1
Traf	2
DK x Posn	1
Dk x ToD	1
DK x Dow	1
Dk x Traf	2
Posn x ToD	1
Posn x DoW	1
Posn x Traf	2
Tod x DoW	1
ToD x Traf	2
DoW x Traf	2
Residual	27
TOTAL	47

Figure 19 -TMA Analysis Of Variance

We have three traffic levels, two positions, two Time of Day two day of week and two teams, making 3 x 2 x 2 x 2 x 2 = 48 values – two from each run.

Now, what about the measures?

Basically, most of all the stuff we did for Sectorisation does not matter here. If we go back to the original Massie grid for Dummkopf (Figure 7) we can cut out the unwanted measures – including Heart rate which the buggers lost. That gives us seventeen measures, with fifteen sources of variation which gives 255 tables for liz – most of these will, I think will not have enough data to make a valid judgement.

Ok – let's do it.

And – Wendy, you can collate the end of simulation questionnaires

Petra – you do the TLX graphics as Liz asks for them..

Measure	DUMMKOPF ON / OFF	Comment
No.Land	=	Expected
No. Dept	=	Expected
DelayLand	-	Bad
DelayDept	-	Bad
No.Orders	=	Expected
Tim.Orders	=	Expected
Pct Orders	=	Expected
FailOrders	-	Bad
Value ISA	- -	Bad
Time ISA	=	Expected
NoMisApp	=	Poor Data
TLXMent	- -	Bad
TLXPhys	=	Expected
TLXtemp	- -	Bad
TLXPerf	=	Expected
TLXEffort	- -	Bad
TLXFrust	- -	Bad

- - = ON much higher (<0.1%)   - = ON higher (<1%)
= = No Sig difference
+ = OFF higher (<1%)   ++ = OFF much higher (<0.1%)

Figure 20 - TMA Measures

Multiple Regression Analysis

The next most commonly used method, after analysis of variance, is Multiple Regression. This method essentially tries to explain one variable as a linear combination of other variables. It derives a set of weightings for the explanatory variables which produces the best fit to the variable explained. In its basic form, it assumes that the variable being explained is a simple weighted sum of the explanatory variables. This may miss more complex relationships. (Analysis of variance is good for testing interactions, but requires balanced experimental designs.) It is perfectly acceptable to apply mathematical transforms to the variables to produce a better fit, provided this is done on rational grounds, and not as a simple shotgun process. For example, taking the logarithms of the variables can produce a multiplicative model, which may reflect biological processes rather better. 'Lag correlation' uses values of predictors at an earlier time to predict the current value of the predicted variable. (It is considered bad form to use later values of the predictors to predict the current value of the predicted variable.) Here again, a competent statistician should be consulted. Recourse to copying a previous researcher, computer package salesmen, their literature or the Internet is very inadvisable.

A multiple regression, however derived, will usually suggest that some of the explanatory predictors make no significant contribution to the predictor equation. These may be discarded, with caution. Sometimes predictors are closely correlated, and it may be difficult to determine which ones to discard. There are different methods available, such as starting with one predictor and adding others in succession or starting with all predictors and discarding them one-by-one.

When a satisfactory prediction equation has been obtained, the residual differences between the prediction and the predicted value should be examined, graphically and statistically to identify any systematic errors, which may suggest unmeasured variables, and non-normality of the distribution of residuals, which may affect the statistical significance of the prediction equation. Many streams of experimental data show a significant time-dependence. For example ISA values usually start at a low value at the beginning of an exercise, build up to full activity and tail away as the simulation 'runs down'. In traditional Air Traffic Control the work of the Planner, which concerns aircraft about to enter the area, peaks considerably earlier than that of the Executive, who is generally concerned with aircraft in or about to leave the area.

Since many potentially important variables vary on different time scales, and come from different sources –questionnaires, on-line recordings, system records, biological measures, it is advisable to have interpolation and synchronisation programs available and well-tested before carrying out the analyses. Some computer statistics packages do not concern themselves with this aspect, some provide complex methods, and some rely on the manual efforts of their unlucky users.

Dopey; Christ, I've got enough to do as it is if Liz wants any of this she can bloody well ask for it.

Factor Analysis

Factor analysis differs from multiple regression in two major aspects. There is no single variable that is being predicted, and the technique is descriptive, not analytical. The underlying theory is laborious, if not complex. The method forms a matrix of the correlations between all the variables available, then tries to extract factors which explain the variation in the most economical way. The principal component is a linear combination of variables that explains as much variation as possible. The second component explains as much as possible of the remaining variation and so on. It is usually a matter of judgement in deciding how many factors to retain. If there are as many variables as there are sets of data, the mathematics involved mean that there is an exact solution to the equations. If there are more variables than there are data sets, or if some variables are linear combinations of others, the mathematics break down, and the method cannot produce a meaningful result. (Some statistical packages do not warn the user that this is happening.) Although factor analysis itself does not determine the statistical significance of the factors produced, an idea of their relative significance can be made by subdividing the data set into two, randomly allocating values to each, and comparing the resultant sets of factors. Alternatively, small random disturbances can be introduced into the data set before running the factor analysis. If the analysis is repeated (say 100 times) with different random disturbances, and the resultant sets of factors plotted, any substantial factors can be seen.

Dopey; Ditto. if Liz wants any of this she can bloody well ask for it.

Canonical Analysis

Canonical analysis is a cross between multiple regression and factor analysis. A set of predicted and a set of predictor variables are defined, and the analysis tries to find a set of weightings (factors) that predicts as much of the common variation of the predicted variables, then another that predicts as much of the remaining variation as possible, and so on. By this time, the underlying ideas have become so complex that most reasonable people give up and go away.

Dopey; We already know which measures are linked with which. It might be interesting to do this, but it will have to wait.

Automatic Interaction Detection

Automatic (or Artificial) Interaction Detection (originally known as AID, although this acronym has now acquired less attractive overtones) is a computer based technique. Although there are many different versions, the commonest involves specifying a predicted variable, which may be on any scale of measurement, Nominal, Ordinal or Interval, and one or more predictor variables, which may also be on any scale of measurement. The computer program then tests each possible partition of the predicted variable according to the values of the predictor variables for the statistical significance of the difference in the partitioned variable. For example, the predicted variable might be the NASA-TLX score of the operator for a specific exercise, and the predictors might be the type of working position, sector of operation, gender and age of the operator. The program would compare the values for the TLX score for each type of working position with the TLX scores for all the others, and retain the most statistically significant difference. It would compare each sector with all the others in the same way, then the different genders, and those above and below each class level for the range of age classes. It would retain the most significant split overall. It would then select (depending on the user's choice and the scale of the predictor) the partition with either the largest number of data, or the largest variation within the data. It would repeat the process, finding the most significant split of the predicted variable. Depending on the user's choice, it may try re-combining partitions to maximise the variation in the predicted variable to be accounted for by the least number of splits. It may do this after splitting up the predicted variable into the maximum number of classes, or after each split, again at the user's discretion.

Although this technique can provide some interesting insights, particularly where there is a very large number of predictors, it is has some drawbacks. The 'probabilities' reported by the tests employed are not valid significance estimates, because a very large number of tests is carried out, and some 'generated significance' is bound to occur. Where predictor variables are closely allied, the choice may be practically random which predictor defines the partition. Moreover, some of the partitions report statistical differences that are very large in terms of conventional tests, and may produce associated probabilities that are too small to measure by conventional computer routines, which report them as zero, making comparisons impossible. Sometimes the statistical differences themselves can be compared, although the logical processes involved can be misleading.

-- -- ---

Dopey; As above..

Fancy Analysis

There are at least as many methods of statistical analysis as there are statisticians – possibly more. Many of these are extremely specialised (both methods and statisticians). If there is an analytic method in general use in your field, by all means use it. If a method is not in general use, but you feel it can be useful, use it, but do not expect the readers of your report to know what it means. Senior management did not get there by being blinded with science. Strangely, they may not trust 'the experts' – especially if they do not like what they are being told.

Dopey; As above.. In Spades!

Postscript

If you have not been discouraged by all this, you may benefit from reading Kendall (1973, 1975). For nonparametric statistics Siegel(1956).is a valuable guide for the confused. These books, written before the 'computer revolution' give considerable detail on how the various tests are calculated, which may not be available with contemporary proprietary software packages.

Dopey; And the best of British Luck!

Statistical Description

Once statistically significant differences have been found, they must be reported. Apart from a few terms, such as 'mean' and possibly 'standard deviation', the terms used by statisticians are not understood by most potential readers. Graphic images can provide extremely useful means of communication, if properly designed and ethically employed.

Because most statistical analyses concentrate on detecting and assessing statistically significant differences, statisticians tend to forget that a statistically significant difference may not be large enough to have practical importance, or may be of no practical interest in any case. If, for example, analysis shows that red-headed people make very poor controllers, but policy prohibits discrimination on the grounds of hair colour, the information is of no practical use. Equally, non-statisticians are often tempted to base conclusions on differences that are not statistically significant. This, to the classically trained statistician. is simply wrong. In real life, a finding that there is a 75% chance that A is better than B can be useful if the client must make a choice. Bayesian statisticians revel in this sort of thing, although most Bayesian statistical analyses seem to come out the same way as their frequentist equivalents.

Tables, Figures and Diagrams

We are all familiar with a variety of tables, figures and diagrams. Some of these provide useful information in an easily digestible form, some provide useful information in a difficult form, and some are, intentionally or not, misleading. Unfortunately, the design of diagrams is often left to 'designers' most of whom have little interest in providing exact information, and will always prefer a 'memorable image' to a useful one.

Yau (2011) is a particularly valuable resource , since he gives practical guidance on how to generate diagrams, tables and other images using both commercial and free software. He also gives valuable advice, with examples, on how to lay out these images.

Some traditional displays are inherently misleading. For example, the 'pie' chart, particularly in its more fanciful versions, involving colour, shading and slices extracted, can be shown to give a wrong impression of proportions. Judging the relative sizes of different slices of different sized pies is practically impossible, which may explain their popularity with some political parties. See Bertin (Bertin 2011), who, being French, naturally refers to Camemberts rather than pies.

Two- or three-dimensional images look good, but can be very deceptive. If the second of two items is actually twice the first, a two-dimensional image will exaggerate the difference, and a three-dimensional one will be even worse. If the quantity is represented by the volume, rather than linear measure, it will be underestimated. If the area of a three dimensional image is used, anything may happen.

Tables

In about 1500 BC scribes were writing on clay tablets at Knossos (Chadwick, 1958). What they wrote was not poetry, prayer or even correspondence, but tables of data. As a means of reporting data, tables have by far the longest history. Microsoft Excel is a spiritual descendant of these tables. If you have a large number of data to display, tables are the simplest way to do so. If you wish to show, for example, at what time a particular train will arrive at a station, then a table is a good way of doing so, since most users will be looking for exact data on specific items. If you are looking for the biggest or smallest value, or are interested in how several different measures vary, together or separately, then tables are not the right way to display the data. If you are monitoring events in real time tables are definitely not the right way. When Apollo 13 lost half its oxygen supply the first thing the monitors on the ground knew was when the astronauts told them "Houston, we have a problem". In fact, a value on a tabular display had changed from 100 to 0, but this was one value among several hundred, and it was practically impossible to notice the change.

If you do have to use tables there are some rules that you should apply:

- Keep it simple
- Label the table
- Arrange the rows and columns to suit your readership
- If you expect the user to follow a line of the table, make sure alternate lines are backed by a light colour, or use horizontal lines on some logical basis to guide the eye.

Bar Charts

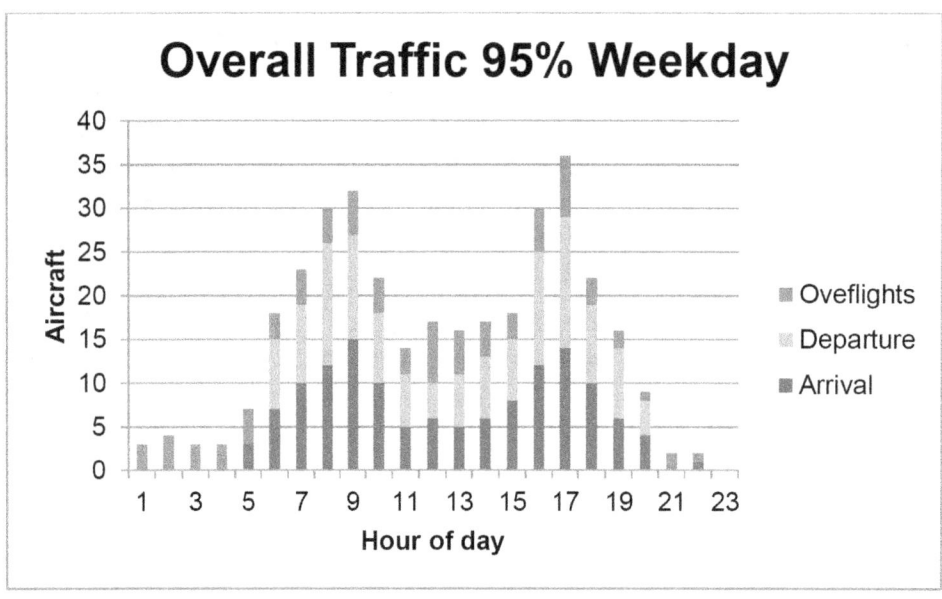

Figure 21 – Overall Traffic 95% Weekday

The traditional bar chart is probably the best known diagram form.

(Figure 21 which reproduces Figure A2-2 in the report given in Appendix 2, is a useful example.)

Even so, there are still pitfalls. When comparing quantities, it is often possible to incorporate sub-divisions of the total quantity in each bar. Care is needed in arranging the order of sub-divisions, placing the least quantities at the bottom of the bar. The order of bars may be determined the situation.. Here, the order of bars is determined by the time of day. Except where readers may be expected to be preferentially interested in one column, alphabetical order is rarely useful. Increasing or decreasing order of overall size is most often useful. The bars have three levels of shading. These are in black and white. It is tempting to use coloured bars, but many documents still do not appear in colour. Cross-hatchings, particularly diagonal, can give optical illusions, not a desirable effect. More than five or six subdivisions become confusing.

In this diagram, the identities of the columns are intuitively clear. Where the columns need long names, it may be preferable to turn the chart on its side, using horizontal bars to allow space for the column titles.

Line Charts

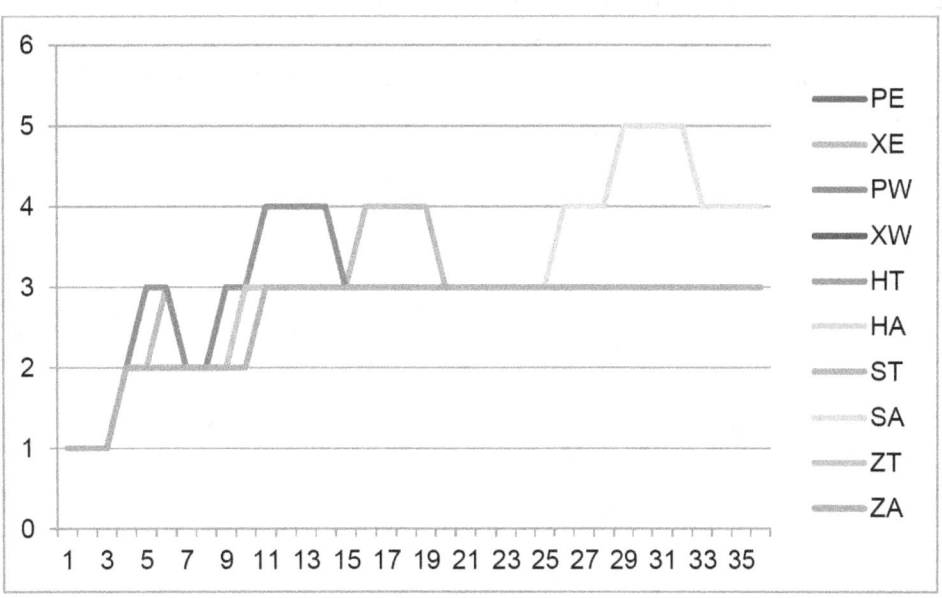

Figure 22 - ISA Values 15th 0900 2 Sector W/E 100% Traffic

Line diagrams are particularly useful to indicate the way in which a measurement varies with time. It is not usually worthwhile plotting more than six to eight lines on the same diagram. The diagram given as Figure 22 is not perfect. For example the axes are not labelled, and the scale along the bottom (the X axis) gives the number of the reading in order, rather than the elapsed time. Equally the unlabelled Y axis which is in fact the number from 1 to 5 corresponding to the key pressed does not have either an axis label or a label for each level. It also includes scale levels of 0 and 6, which cannot occur. The different shaded lines correspond to the ten controllers in this exercise. Most of the values are overwritten by the last (ZA – Zenda Approach), but the diagram does show that the West Planner (PW), the East Executive (XE) and particularly Strelsau Approach (SA) had difficulties. (Normally, these lines would be coloured, but they can be presented in monochrome – this is done here because it costs five times as much to print in colour.)

This is a diagram taken directly from Microsoft EXCEL, and shows how it can be used to display data without the need to print it out. This sort of diagram can be used to identify what is going on in vast complex files of data, created and discarded in minutes. If a diagram is to be presented in a report, time should be spent on improving the layout and visibility. Here again Yau (2011) is a valuable guide.

Nightingale Diagrams

Nightingale diagrams, sometimes called 'Spider Webs', but properly named after Florence Nightingale, who invented them, (Nightingale, 1858) in Yau (2013), can be used for measurements on a cyclic basis, such as deaths in military hospitals by month, as she used them. They can also be used where you have a set of independent scales, such as are provided by the NASA-TLX.

Figure 23 is an example of a NASA-TLX summary, drawn from the report in Appendix 2, (on a larger scale simply because we have more room here).

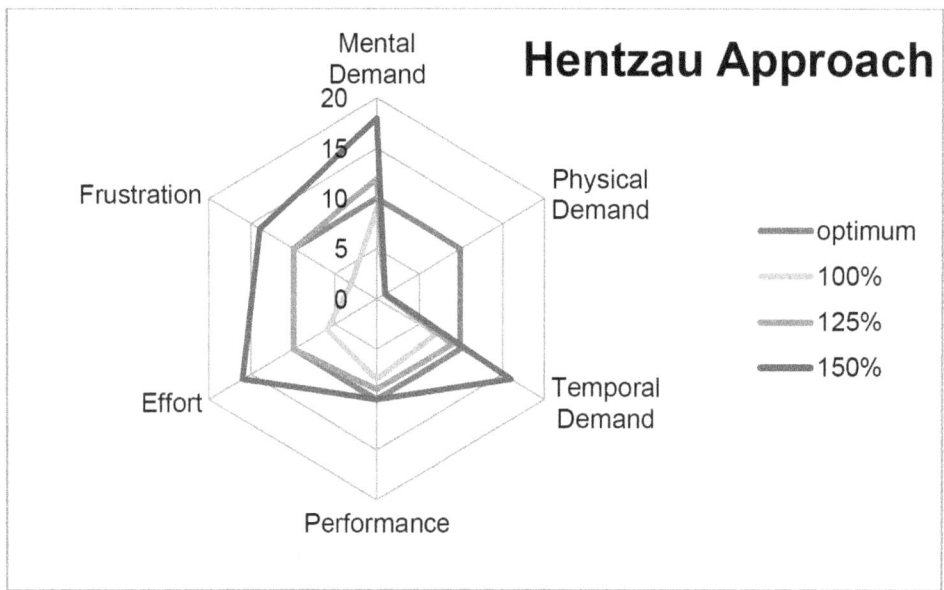

FIGURE 23 - NASA TLX components

Here the six axes correspond to the six scales of the NASA-TLX. The regular polygon is the mid-point of each scale, representing an optimum use of human resources. The 100%.125% and 150% polygons represent the average responses for traffic samples of the corresponding percentages of the current 95th percentile traffic.

It can be seen that there is a regular progress from 100% to 150%, except that the physical effort is negligible throughout. The 125% polygon is close to the optimum. The 150% polygon shows heavy mental effort and temporal demand and substantial 'effort' and frustration.

(In practice, although the six scales are conceptually independent, they are usually statistically correlated, so it is rare for polygons to cross.)

Where colour is available, these lines are best distinguished by different colours. If colour is not available, dotted or dashed lines would be preferable.

Maps

Most reports contain at least one map, usually describing an area that is being investigated. The classic, if dated works by Bertin (1981; 2011) tells you more than you need to know about the design of maps. (Bertin was a cartographer, working before the Internet and computer graphic developments.) Yau (2013) gives a more modern discussion, with practical hints on how to access internet resources.

In general, it is wise to avoid 'over-egging the pudding'. Include only what is needed in the context, because irrelevant detail may distract the reader. Use faint tints to colour large areas, and vivid colours for point features.

Figure 1 (in Chapter 7) is a typical map for a report. Make the map as large as is practical, since a small-scale map will usually have illegible text and unrecognisable symbols. Give the scale as an annotated bar, not as a ratio (1 cm = 1 KM) since the size of the map may change during reproduction or copying.

Chapter 14 – Reporting

This chapter gives a short guide to the presentation of the results of a simulation – as a technical report, as a 'conference paper', as a verbal presentation and as a poster. If you submit a conference paper and it is accepted, you may be required to present it as a verbal presentation, or as a poster. Different audiences may require presentations adapted to their interests. They have in common that the presenter must attract and keep the attention of his audience. The mere fact that what the speaker, or writer, is saying may change their entire future will not stop them falling asleep on a hot afternoon after a good lunch.

Technical Reports (Appendix 2)

The main permanent record of a simulation is usually a technical report. Most organisations have a 'house style' which serves as a guide as well as a trademark. Where contributors come from different linguistic and cultural traditions, it may take a considerable time – and management support - to establish such a style. I produced a draft style for the Eurocontrol Experimental Centre six months after I joined, in 1970, and was still trying to get it formally accepted on my retirement in 2002. (Other simulation centres were in fact using it – how they got hold of it I shall never know.)

Appendix 2 is a complete technical report for the dummy simulation "Ruritania I" described in the previous chapters. It may be instructive to compare the report with the narrative embedded in Chapters 5 to 13. The general form of this report is characteristic of such reports. They usually begin with a number of technical pages, essentially for accounting purposes. It is common nowadays to include a page detailing when the report was issued, and when and how it has been modified. Technical reports are usually filed and forgotten, so they are virtually never modified. However, there exists a standard for documentation, which requires such a page to conform to the standard

-- -- ---

Liz; OK – No more Sun and Fun. Back to work. Usual Boiler Plate to start with -

The 'Executive Summary' is a short description of what the report says, provided for the lazy and ignorant (busy and wide-ranging) who need to appear to know what it is about.

Liz; I'll put this in last – when we know what the results were.

Chapter 1 provides the background to the simulation, and describes the layout of the following chapters.

Liz; Easy -. Basic intro and the standard template.

Chapter 2 lists the questions raised.

Liz; Copied straight from the simulation definition in chapter 6.

Chapter 3 provides a brief description of the relevant airspace and traffic, or the equivalent in other fields of study..

Liz; Chapter 7 figures 1, 2 and 3 for the airspace, and Rupert's pretty pictures for the traffic..

Chapter 4 describes the experimental plan, with the proposed measurements.

Liz; Chapter 8 figure 5, with a detailed explanation of the exercise coding.. plus a text note of the various failures, for the experimental plan. Usual boilerplate for ISA and NASA-TLX. . List the questionnaires - no room to put them all in.. Massie grids for Sectors and TMAs with explanation for system measurements. Better leave out Heart Rate – it never got off the ground.

Chapter 5 describes the running of the simulation.

Liz; Surprisingly short – mainly because there were no real problems. Must not forget the visitors.

Chapter 6 summarises and discusses the analysis.

Liz; This is the sticky one. We already know the answers, but we need to show them, - I will have a talk with Dopey – he is good on graphics.

Begin with the usual boilerplate about subjective and objective measures. Pads it out a bit - no that's in Chapter 4.

To start with – One sector versus two? We really only need the RU versus RE and RW – the TMAs shouldn't matter for this. I had better have a word with Dopey.

Now DUMMKOPF – We only need to look at Strelsau International – and really only Approach for this. We may have some objective differences, but it is the ISA and NASA-TLX that matter here.

So Hentzau and Zenda do not really matter, except to show how their load will increase with heavier traffic. They are neither near saturation. We will have to put something in about them, or someone will be asking why they were there in the first place.

(later) Bloody Hell, look at the figures for Hentzau and Zenda weekends! Why didn't we notice that?

Wendy; I did try to tell you, but you told me to piss off!

Liz; So what can we say?

Wendy; A lot of the Hentzau traffic is from the flying school. We could get them to move.

Liz; Or simply not fly at peak times. Who wants to fly at weekends?

Dopey; Most people, it seems.

Liz; tough! They can run away and play somewhere else.. Back to business.

For the sectors, we do ANOVARs, using your clever trick to separate out One/Two sectors and RUE/RUW. We do separate analyses for Weekdays and Weekends. We use the mean values for three-minute measures – from 15 minutes to end of simulation. We haven't got time for peak values. We do ANOVARS for sector level variates, like number of a/c entering and for working position variates – like ISA or frequency. We had better split them up for reporting, communications vs the rest.

For the TMAs we do ANOVARs for weekdays and weekends separately and for the TMAs separately, with TMA level variates and Working position level variates. For Strelsau we have to include DUMMKOPF, not for the others.

In the report, we summarise the significant findings, and provide graphics to show the effects. Using graphics only makes it easier to read - or less bloody difficult. Most managers like to look at the pretty pictures.

Can you do that for the end of the week?

Dopey; End of the year more like. I have Kennaquaire coming up.

Liz; Get Wendy to do the donkeywork. I'll have to do the write-up. It won't go away if you leave it.

Chapter 7 repeats the questions raised, with the conclusions drawn.

Liz; First bit is easy – copy chapter 2 -. The extra conclusions need careful wording..

The report finishes with acknowledgements.

Liz; Wendy, where is the attendance list?
Should we acknowledge the Strelsau Fire Brigade for their rescue of Our Rupert?
Better not – let sleeping dogs lie – or dirty dogs stand.

Now turn to Appendix 2, read the bastard, and compare it with what actually went on in chapters 5 to 14. *(The pig is fiction, the timer, alas, is not.)*

Liz: That will have to do. So we have left out an awful lot -we could have put in lots of tables, and given summaries of the NASA TLX responses and so on, but we just do not have the time.

I will do the conference paper, verbal presentation and poster versions as they are needed

Oh, Sod - the Executive Summary! "In time and on budget" of course.

Conference and Scientific Papers (Appendix 3)

Very few simulations merit a full-dress scientific paper. Scientific papers require a considerable quantity of 'scientific apparatus' such as literature searches, which are beyond the reach or interest of the organisers of many simulations. They also require a considerably higher standard of statistical analysis. Considerable patience is needed, as is a good memory, since you will probably be several simulations further on before you find yourself arguing with referees and editors.

Conference papers are rather different. A typical professional conference involves three to five days of presentations. Each full day has four sessions, each about two hours long, in each of which about six presentations are given. In theory, each presentation usually has a five minute introduction, a fifteen minute presentation and ten minutes for questions. (The guide to verbal presentations given below applies here, but it may need modification to suit the skills and interests of the delegates.) Depending on the conference, there may be one or two half days when the delegates attend study tours, a professional society annual meeting is held, or some similar junket takes place. "Workshops", where members may update their skills occupy some sessions. Theoretically, workshops should involve audience participation and discussion, but frequently they are merely extended lectures. Small societies may have only one session, which allows most of the delegates to attend most of the presentations, medium sized ones may have three or four parallel sessions. Large societies may have as many as thirty separate parallel sessions – in some of which the speakers may outnumber the audience. Most societies hold plenary sessions, where an eminent authority spends thirty minutes addressing all those delegates who have finished breakfast in time and have not contrived to escape, or may even be interested in the topic. It is an unfortunate tradition that invited speakers are not expected to prepare a paper for the proceedings. Certain notorious individuals are generally believed to prepare their address on the plane coming in – if at all.

To complement these delights, delegates are issued with a book of "conference proceedings". This is a semi-professionally produced book, complete with publisher and ISBN, usually an annual dated series. It is the only public record of much research, and the only permanent publicly available one. It is usually handed out on arrival, and most delegates skim through it, looking for their own papers, and references to their work by other authors. (Well-organised conferences have subject and author indexes to their proceedings.) Few delegates have the speed reading skill to read the complete book before the conference starts, and most lack the endurance to read it afterwards. It is consigned to their bookshelves, and may remain untouched for the rest of the delegate's career. In a recent depressing development, an on-line version of the proceedings may be substituted for the printed version, presenting a practically nil chance of papers ever being seen again. (On the other hand, electronic publishing allows more freedom than the classic conference paper described here. – extra pages and colour plates cost nothing on the Internet.)

A potential solution to this problem may lie in the 'RESEARCHGATE' on-line system, where reports may be posted by their authors, and can be read by other researchers. The managers of the site post work from a large variety of journals, dating back many years. (I was interested to be reminded of several papers of which I was co-author in my larval – postgraduate - stage, fifty years ago.) RESEARCHGATE has several interesting problems, which discretion, and an eye to the law of libel, suggests should not be mentioned here.

If a conference involves 120 presentations, the available space in the proceedings must be strictly rationed. Five pages is a normal allocation. Since the paper is required in 'camera-ready' format, the format, typeface, font, justification, margins and so on will be strictly defined by the appointed editors. One common typeface is Times New Roman, usually in 11 point. (A 'point' is $1/72^{nd}$ inch or about 0.35 mm.)

The first page usually presents the title, the author(s) and their affiliations, preferably including a postal or e-mail address, and an abstract, usually limited to about one hundred words. Remaining space starts with an introduction. Many editors require that the subsequent sections should be titled; Method, Results, Discussion, Conclusions and References. (Somehow, the references usually extend to the last line on Page 5.) This suggests either a rigid mind set or unfamiliarity with modern scientific research.

The text is necessarily condensed. Many authors resort to acronyms for each repeated phrase. The acronym is defined the first time it is used, and the hapless reader will spend some time switching back and forth finding definitions. In particular, the Results section is usually riddled with, so to speak, "local acronyms" and conventional statistical abbreviations. For instance: -

"There was no difference between the number of correct responses in the Manual and ACC conditions ($F_{1, 8}$ = 0.079, p =0.786), however there was a significant MWL reduction with AS ($F_{1, 8}$ = 43.566, p < 0.001) and a further reduction when using ACC+AS ($F_{1, 8}$ = 159.599, p < 0.001)."

If you can understand that, you are probably the author, in which case I apologise, and have deliberately avoiding giving a reference. This is a perfectly normal section of prose from a conference paper, and considerably more fluent than many. I have at least one example containing an entire page of this style of communication, without even a paragraph break to draw breath.

It may be possible to avoid some of this by the judicious use of graphs, charts or diagrams. In general, the design of these is discussed above, in Chapter 13. In conference papers, there are two additional considerations.

You will usually be required to submit your paper on the equivalent of A4 sized paper (297 mm x 210 mm). However, this is reduced during printing to about 230 mm by 160 mm. The actual available area of a page is about 170 mm by 110 mm, allowing for the strictly enforced margins. Full-width images are therefore about 11 cm by 6 cm, and, with explanatory text, will each occupy half a page. It is therefore rare to be able to include more than four images in a paper. It is tempting to include detailed captions in tiny print, which will be completely illegible in the proceedings. Print out your paper at the 'two pages per sheet' size option as a precaution.

If it is illegible, remember that this is how it may appear to your older readers – who may be the ones deciding your future career.

A recent template for conference papers includes the following helpful advice.

> "In the text, write out numbers nine or less except as part of a date, a fraction or decimal, a percentage, or a unit of measurement. Use Arabic numbers for those larger than nine, except as the first word of a sentence. .Use abbreviations of the customary units of measurement only when they are preceded by a number, for example, "5 min" but "several minutes". Write "percent" as one word, except when used with a number, for example, "several percent" but "25%"."

Modern technology allows authors to embed images directly in their text, and to 'flow' the text around tall, thin images where these are required. Most conference proceedings are printed in monochrome black and white, although shades of grey may be used. The inclusion of coloured photographs and graphs is sometimes permitted, usually when the conference has a major commercial sponsor covering the cost. Cynics regard abundant coloured illustrations in printed conference proceedings as a danger signal.

There should be a fair number of people interested in the outcome of a simulation. Most of these people will be familiar with the technical language of the field being simulated. It may be wise to 'remind' them of some of the finer points that are relevant to this particular simulation, without suggesting that any specific auditor does not know them.

Verbal Presentations (Appendix 4)

Practically, a PowerPoint presentation provides a useful skeleton for a verbal presentation, subject to the overriding condition that the presenter must NEVER read the PowerPoint slides as they appear. This interferes with the audience's process of absorbing information, and creates the impression that the speaker does not really know what he is talking about. If you have time, do not create PowerPoint slides by copying chunks of prose from reports or other documents. (Ideally, the slides should provide illustrations, rather than text.) In practice, you will usually have to scramble together what is available, making it look as if you knew it all the time. Audiences can read faster than you can speak, and will become distracted from what you are saying. Before you undertake the presentation, make a paper copy of the slides and staple it in order. Well before giving the presentation, go through the slides on paper, checking for spelling errors and missing items. It may be a good idea to make a number of copies of the slides, and tell the audience they will be distributed after the presentation, relieving them of the need to take notes. Do not distribute notes before speaking, since this gives the audience the chance to get ahead of you.

It is usual to start with a slide giving the title of your talk, the name and date of the simulation you are reporting, and the name and affiliation of the presenter. If this is projected at the start of your presentation, it helps to establish to the audience that they are in the right room, and avoids an exodus as you start to speak. It also makes sure that you have the right set of slides, not someone else's. (It has happened!) Some organisations insist that their name and logo appear on every slide, producing a background hash that is effortlessly ignored by all present. If you can get away with it put the extra brushwood on the first and last slides only.

Begin by stating the aims of the simulation. If you are using an up-to-date version of PowerPoint, it should be possible to bring these up one-by-one on a single slide, as you state them (using your crib sheet.) If you have to use an obsolete version, put each objective on a separate slide.

Next, you can describe the physical layout, showing a plan or photograph of the simulated control room, and possible a typical set of working positions. (It is not necessary to go into the dummy and simulator pilot positions.) This will not actually serve any purpose, but will give some people a feeling that they understand the situation.

Describe the working materials – traffic and how it was derived, the simulated space and its divisions, and so on. Here images can be invaluable, including graphics – as described in Chapter 13.

Define the different organisations being tested. If you have not already done so, give the organisations memorable names to ease the burden of assimilating information.

Describe very briefly the running of the simulation.

Then show the aims, one by one, bringing up the response to each – one aim and response per slide.

Bring up any additional conclusions. There will always be additional conclusions, and they are often the most interesting.

Finish with an acknowledgement of the work of the participants, staff, developers and anyone else it may be expedient to thank – the sponsors, for instance.

Use a final slide to call for questions. (It is wise to ensure that the chair of the meeting is primed with at least one question.)

Finally, unless you are a very experienced presenter, and even then, occasionally, make the presentation beforehand to a group of colleagues, preferably friendly, who may be interested in the content and can advise the speaker where his message is not clear, and what he may have omitted. It is not usually sufficient to run through the presentation silently in your office, since you will always underestimate the time. Equally, giving the presentation at full voice to an empty office, particularly if you include rehearsing the PowerPoint slide show, and if the office walls are as thin as they usually are, may give your neighbours the impression that you are mentally disturbed. Whether this is true or not, it is usually undesirable to give that impression.

One unacknowledged virtue of the verbal presentation is that you may include 'spontaneous' off-the-record comments that may be very valuable to particular audiences, but would cause great offence in any official document. For example, a description of the mandarins of Her Majesty's Treasury will generate a hum of appreciation in any scientific audience, but fury in management, and glee if it gets into the hands of the 'Daily Mail'.

Appendix 4 is a verbal presentation of the Technical Report presented in Appendix 2. This is the version prepared by the author, with her prompts. It would need to be cleaned up before providing print-outs.

Posters (Appendix 5)

You may find that a paper for a conference is accepted, but that you are asked to present it as a poster. You will then have to present your information in a primarily visual form. This can be very wearing, since it involves transforming a serial presentation to a parallel one, requiring considerable mental effort if it is to be done properly. Resist the temptation to simply copy and paste parts of your paper. Often a consideration of the actual content of a paper will show that there are parts that are parallel developments. It may be tempting to make a bonfire of the poster after its single exhibition at the conference, but it can be more useful to frame it and put it on an office wall.

The poster size will usually be specified by the conference organisers. You can produce a poster using Microsoft WORD, and using the A3 standard size (297 x 420 mm). Even if you can only print up to A4 size (210 x 297), high street printers can usually produce the poster as A3 (297 x 420 mm), A2 (420 x 594 mm), A1 (594 x 841 mm) or A0 (41 x 1189 mm), from a ".docx" file . It is as well to find out beforehand whether the poster should be 'portrait' (taller than wide) or 'Landscape' (wider than tall). If you are showing a poster, it is easy to print a number of A4 size copies. (If the A4 size copy is illegible, then the type size you are using for your poster is too small). Paper-clip or staple these copies to your poster, so that visitors can take them.

Appendix 5 is a poster representation of the conference paper shown in Appendix 3

Website postings

A website posting is today one of the most effective means of distributing information. If your organisation, or you yourself, have a website, it is relatively easy to post a PDF version of the technical report. If you have time and effort available, it may be worth modifying the technical report to include additional images and graphics – colour comes free on the Internet. You may incorporate any relevant PowerPoint 'slides'.

It is unwise to post a Microsoft WORD version (.doc or .docx) since these may easily be tampered with, leading to some embarrassment, and possibly unwanted publicity.

Social Media

It may be worth posting a short non-technical description on whichever social medium is currently fashionable. Be very careful – "here be trolls!"

Chapter 15 – Future Developments

General

The practice of control room simulation will become increasingly popular as computer-based supervisory techniques continue to replace traditional routine manual and mechanical handling methods. In general, control is passing from direct routine operations to intervention only when needed. This will lead to 'loss of the picture' – as Air Traffic Controllers call it. If you have ten aircraft in your sector, you can probably remember most of the relevant information, although you will forget information that seems irrelevant at the time. You will be haunted by a justified fear that you may be missing something relevant, simply because it seemed irrelevant at the time it appeared. If you have fifty aircraft in your sector, you will simply not be able to cope, and will be obliged to rely on luck and safety margins. You will know it, and it will destroy you.

Supervisory control differs from traditional 'hands-on' control. The supervisory controller is alerted by the system when routine procedures break down. (S)he must grasp the situation and react in a non-routine manner. This will require constant refresher training. I suspect that many future control rooms will be constructed as twins, one of which is normally used for real control while the other runs simulations. Control teams will spend a substantial proportion of their working time in simulation, reviewing 'incidents', arguing who was responsible, practicing emergency drills and carrying out refresher training.

In the longer term, Control Room Simulation may well follow occupations such as typewriting and car driving, becoming so generally used that it ceases to be a specialist trade, and is accepted as part of the routine of control room operation.

At the same time, control rooms are becoming increasingly generic, adapted to the normal range of human ability, and relatively undifferentiated. This will make the 'general purpose simulator' operated by a specialist contractor, independent of any particular enterprise, viable in itself. Such simulators will be able to simulate many different types of control room. Although my experience is practically limited to Air Traffic Control simulators, the same simulator could be used for aircraft and air systems, trains and railway systems, ships and seaways, road traffic, including tunnels and bridges, electricity generation and distribution networks, gas extraction and distribution systems, chemical plants and pipelines, fire, police and ambulance services and many other control rooms The International Standards Organisation has produced a seven-part standard (ISO 11064-1 to ISO 11064-7, 1999-2008) defining control rooms in detail. SBFI (2015) is a catalogue of ISO-compatible control room equipment from a prominent manufacturer. It is more succinct, and considerably cheaper than the complete ISO standard.

A general-purpose simulator will be used where a 'one-off' simulation is required, and there is no expectation of further simulations. It will be also used for exploratory development where an existing control room is unavailable, or where there is a temporary demand for additional simulator capacity. It may even be used as a forensic tool in reconstructing catastrophic or culpable series of events in judicial and legal contexts. Finally, It will be used as a sales tool for control room builders

The present situation

Figure 24 - Contemporary Large Real-Time Simulator

To illustrate the current situation, Figure 24 shows the EUROCONTROL Real-Time ATC Simulator. Depending on the simulation, it may have up to 40 working positions, with 20 simulation staff, showing 200 aircraft simultaneously, in up to 20 sectors. Each run is from 60 to 90 minutes, including a running-up period, and a simulation may involve up to 4 runs per day for five days a week for up to four weeks.

It is mainly used for -

- Validation – will a proposed system actually work?
- Modification of existing systems – splitting sectors or re-aligning them.
- Acceptance trials – will the controllers accept a new system?
- Innovation – is a proposed new tool actually useful?

The problems are : -

- that it is costly, slow and complex to set up, requiring virtually a re-build for each simulation,
- It is limited to about 6 simulations a year.
- It does not usually use complete teams of controllers,

- It has synchronisation problems,
- it often experiences problems with defining its raw material – air traffic
- It has grown by accretions and remodelling, so that only a few operators know what is going on.
- There is no complete set of documentation.

Figure 25 - Contemporary Control Room

Figure 25 shows a typical contemporary control room. Operators normally work in pairs, for good psychological reasons. In addition to the desktop displays, there is usually a single large common display, which may show an overall schematic, a direct view of the process being controlled, or selected CCTV displays. Although many traditional control rooms provide a direct view of the process being controlled, CCTV displays can provide, for example, views of the other side of the process, or fine details of particular points of interest. Furniture is to the ISO 11064 standard.

Future trends in control rooms, which will be reflected in simulation, include: -

- A change from the wall-to-ceiling spread of displays typical of many control rooms, which cut operators off from each other, and usually present more information than any human can absorb.

- Iincreasing modularity of both hardware and software.

- Migration from hands-on control to supervisory control with only occasional intervention.

- Increased integration of data sources.

- Predictive displays, showing what will happen if a proposed order is implemented, will become more common. They may, in fact, incorporate a simulation facility. A primitive simulation study, incorporating a simulated simulation facility was test as long ago as 1978 (David, 1979)

- Warning displays in response to emergencies.

- Displays to help the operator resume direct control from a passive or supervisory mode.

- Artificial Intelligence derived operator-system interfaces will be introduced, on a longer time scale.

Simulation facilities for future control rooms will take advantage of these trends.

It seems probable that there will be two distinct directions of evolution for simulators. One will be towards simulators integrated into actual control room facilities, the other towards general-purpose simulators.

Figure 26 shows the probable overall layout of a future control room, with on-site simulation facilities. Real-world static data describes the world in which control operates. The real world is represented in the control room by data extracted from the real world by various computer-controlled and mediated systems. Much data gathered may be irrelevant to the task in hand, and a data concentrator, specific to the field or unique for the control room will put the mass of data into a form suitable for processing. In the actual control room, the concentrated data will be processed to present the necessary displays by a field-specific display/control interface.

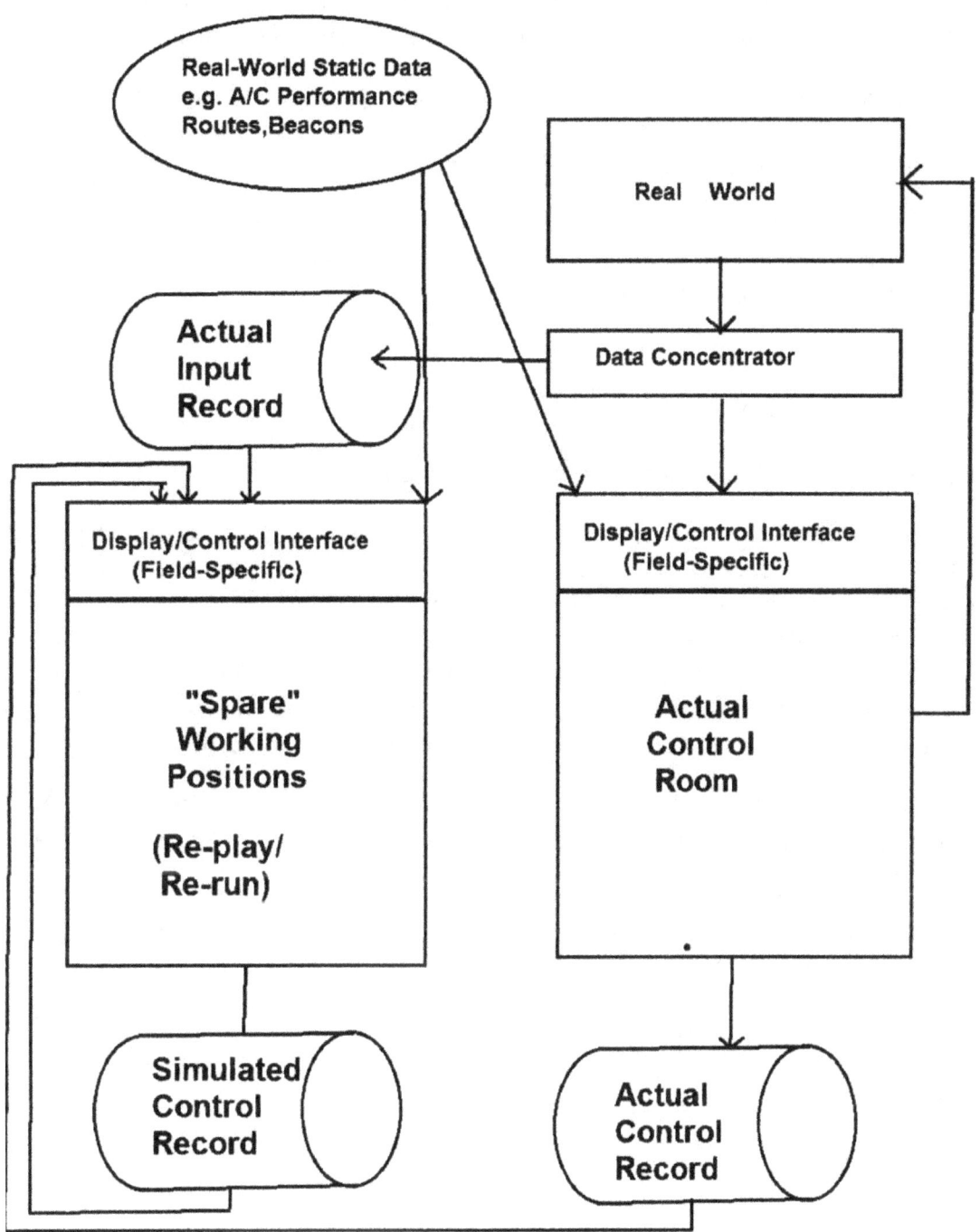

Figure 26 - Future On-Site Simulator

. The actual data will be stored, as will a complete record of all inputs from the controllers. Most of this data will never be used, (like contemporary CCTV recordings) but sometimes it will be vitally important.

On-site simulators will become increasingly necessary as the control mode changes from 'hands-on' to 'supervisory'. They will provide the possibility to replay the control record, or to present the recorded actual input data so that controllers may practice their skills.

These simulators will be used to develop and maintain skills and to practice emergency drills. They will be used in adaptation training and in mentoring. They may also be used for the examination of 'incidents', whatever form these may take

There will probably be between 200 and 500 instances of on-site simulators, using off-the-shelf software, and the hardware of the real control room. Ideally, the on-site simulator will be a replica of the actual control room, and may actually form a complete stand-by control facility.

General Purpose Simulators

Figure 27 shows the general structure of a general purpose simulator.

The relatively few general-purpose simulators (two to five in a country) will be used for many purposes. They may be used for training of complete teams, particularly where the new control room is not yet ready for use. They may be used to validate changes to existing systems. They may be used in development, for example where AI systems are to be introduced, to validate operator-system interfaces. They may be used, sometimes, to examine what may have happened in serious incidents, where the actual control room may be inaccessible. Finally, they may be used to promote sales by more progressive control room equipment firms.

They will include separate 'feed' controllers, where necessary, and 'simulator pilots' or the equivalent in other fields, where control is exercised by giving instructions to the actual operators of vehicles.

It is worth noting that they may work either from input records synthesised from field-specific data, or from input records of actual real-world traffic.

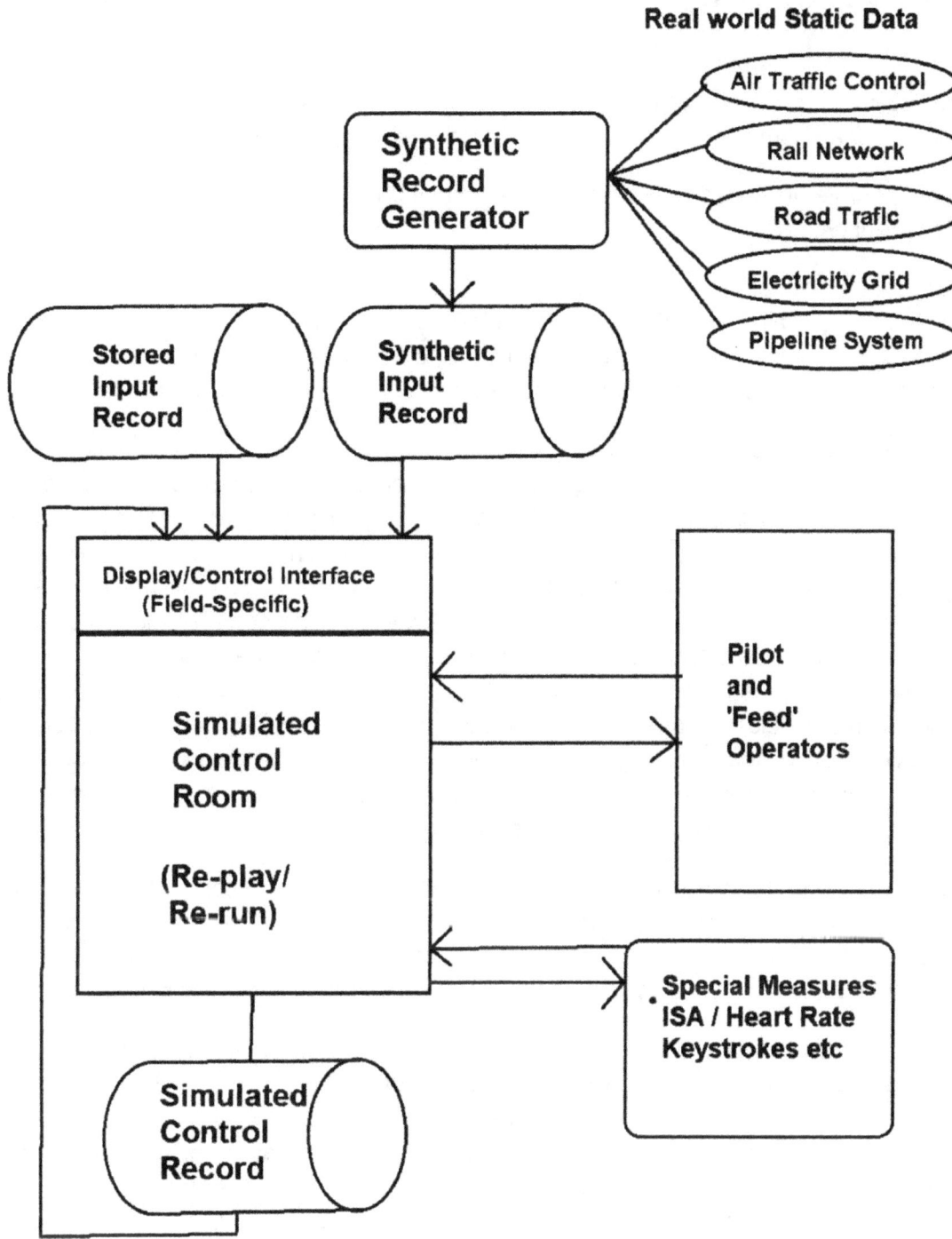

Figure 27 - Future General Purpose Simulator

Chapter 16 – Summary

Real-Time Control Room simulation is not simply a technical system. It is a complex of human-human and human-computer interactions – as is Real-Time Control Room operation.

In this book, I have described how a Real-Time simulator is built, operated, and used to provide answers to real problems.

There are many special purpose simulators designed to simulate vehicles, such as aircraft, trucks, cars and train drivers' cabs. These are occasionally connected to control room simulators, usually where a new type of vehicle is to be introduced into an existing system. There are several good books about these simulators, so they are not described here.

A Real-Time Control Room simulator simulates control rooms. These are increasingly remote from the activities they simulate, for several good reasons, among them safety, economy and accessibility. It is often easier to see what is going on in a process by looking at carefully positioned closed circuit TV than by looking out of a window onto an airport, a rolling mill, a refinery or a nuclear reactor. The actual design of control rooms is not part of this book, since it now well established, international standards are available, and control rooms are available 'off-the-shelf' from several reputable manufacturers. Since they are designed to fit around human beings, rather than to follow mechanical limitations or outworn traditions, they are very similar across many fields of human activity. Usually they consist of rows of working positions, often in pairs, with a common large scale display visible from all working positions, and a supervisors position behind and slightly above the operators. Many control rooms have a visitors' gallery above and behind working positions, preferably separated by a glass screen.

This book describes contemporary Real-Time Control Room simulation, illustrating it with a fictitious example from Air Traffic Control. (A description of 'traditional' en-route ATC simulation is provided, for the benefit of readers in other fields of control, and to provide a background to the description of the planning, running, analysis and reporting of a fictitious simulation.)

It begins by discussing, briefly, the physical structure of simulators, moving on to the people needed to operate a simulator. Usually a Project Leader and an Assistant Project Leader are assigned to a simulation, remaining with it from the initial proposal to the final report - and sometimes after. The other members of staff work with the simulation for periods of about a month around the actual simulation. Specialists are responsible for the overall running of the simulator, the hardware and software, the planning of simulation organisations and runs, data collection from system and participants, analysis of the data, leading the group of 'Feed' controllers, carrying out the actual feeding operations, and running the group of 'simulator pilots' who play the role of pilots in the simulation runs.

It looks at the participants (not subjects) drawn from the field of activity being simulated. These participants usually have a far better and more detailed knowledge of what is really happening in daily routine than those farther from the 'shop floor'. Even where supervisors are drawn from the operators, their knowledge of the problems may well be obsolete, without their being aware of it. Ideally, complete teams or 'watches' should be used, but this is rarely possible. It is advisable to allow for a good deal of 'shake-down' time, so that a basic team develops.

The first step in any simulation begins as the definition of the simulation. This needs careful study. Some examples of good and bad practice are provided.

The data employed in a simulation can be divided into several categories. The most basic data - sometimes called 'static' data - consists of relatively unchanging data. In ATC this includes maps of the relevant area, flight characteristics of relevant aircraft, normal operational procedures and so on.

'Dynamic' data is the traffic that will be controlled. For each run, a sample of traffic is prepared. (In Air Traffic Control, these samples are essentially lists of aircraft call-signs, with an associated type, a route (defined by beacons and airways), a planned start time (either departure from an airport or entry from an adjacent area.) It might be thought that a simple record of relevant traffic would be sufficient, but in practice, there are very few entirely normal periods. Since it is never advisable to repeat a sample, it is better to prepare a very large sample, and select flights randomly to provide traffic at the required density. It is also advisable to have a facility for viewing the sample to ensure that traffic is not already in conflict on entry.

An experimental design can be laid out in advance, but a considerable allowance of 'spare capacity' should be provided, as something will always go wrong. The luckless experimental designer will invariably find himself re-designing the experimental design 'on the run'. The resultant final design will rarely satisfy the statistical purist.

Consideration must be given to the target of the simulation. This may be to assess 'capacity' - always a dangerously vague concept. It may be to assess 'system performance' - assessing the effort required for each unit of output, or identifying bottlenecks in a proposed system. It may be an assessment of 'operator performance', where an operator's performance is compared with a theoretical model, although a large Real-Time Simulator is not the best tool for this type of assessment.

There are many measures of what is happening within a simulation run, 'objective' measures such as the number of aircraft entering an area, or the mean time taken to input an order, or 'subjective' methods, such as the ISA or NASA TLX. More general questionnaires may also be useful. De-briefing' - individually or in groups - is an essential part of measuring what is happening. The simulation leader must be able to distil the essence of operators' comments.

In addition to the hardware and software preparation, care must be taken to show the operators that they are considered to be valuable consultants, not objects of investigation. There must be a well-appointed 'Participants' Room' , showing that pains have been taken to make them feel at home.

The actual running of the simulation is the most intensive part of the task. Accidents will happen, and the supervisor must be ready to make immediate decisions on what to do. Equally, the experimental designer must be able to re-balance the experimental design at a moment's notice.

At the end of the actual simulation, a 'winding-up' and 'thank-you' party is essential, and vital information is often picked up at that point. The simulation leader and assistant must be able to absorb alcohol and information at the same time.

After the exercises have been completed, the data analyst must start to reduce the data to manageable forms. In most cases, she will have to discard much of the data, simply because there is no time to run all the analyses and digest their results.

The analyst will have the simulation leader pressing him for material for the report. There is almost always a report. Very often it will finally be printed long after the simulation has been completed, by which time the conclusions will have been accepted and acted upon, or rejected and quietly filed.

Where time allows, the report may be summarised in a conference report and presented at the conference as a verbal presentation or a poster.

The control room simulation described here is an Air Traffic Control room, but the lessons incorporated are generally applicable to many other fields of human endeavour. Among these are aircraft and air systems, trains and railway systems, ships and seaways, road traffic, including tunnels and bridges, goods distribution, electricity generation and distribution networks, gas extraction and distribution systems, water storage and distribution systems, currency and stock market trading floors, chemical plants and pipelines, fire, police and ambulance services and many other control rooms..

Real-Time Control Room simulation has a secure future in the training, maintenance development and validation of control systems.

It is likely that future control centres will have more than one actual control room, as the mode of control evolves from the present 'hands-on' technology to a more supervisory method of operating. This will lead to considerable changes in what is displayed, and how control is exercised. It will require much more training, particularly to maintain familiarity with the system, and to establish the mutual trust and preparedness of teams of operators. It is quite likely that one control room will control actual processes, while another will be used as a training, development and analysis facility. Considerable economies of effort and increments in safety will be achieved by the modularisation of control systems and the integration of digital control and communication systems into fields where vehicles are currently practically autonomous.

Properly tested 'off-the-shelf' systems will replace the costly and error-prone process of designing simulators from scratch. The human-computer display and data collection software may be specific to particular fields of activity, but data storage, analysis and re-play software may be completely general.

Some general-purpose simulators will be developed to simulate a range of control rooms, for initial development, task allocation and training where the projected system is not operational, and for forensic, sales and other uses.

However, simulation can never become a routine process. It depends entirely on the human element to make it work.

Bibliography

Baddeley, A. D. (1988). A 3 Minute Reasoning Test Based on Grammatical Transformation *Psychonomic Science, 10*(10), 2.

Bertin, J. (1981). *Graphics and Graphic Information Processing* (W. J. Berg & P. Scott, Trans. Second ed.). Berlin: de Groyter.

Bertin, J. (2011). *Semiology of Graphics: Diagrams, Networks, Maps* (W. J. Berg, Trans.). Redlands California USA: esri Press.

Casali, J. G., & Wierwille, W. W. (1984). On the Measurement of Pilot Perceptual Workload: A Comparison of Assessment Techniques addressing Sensitivity and Intrusion Issues. *Ergonomics,, 27*(10), 50?

Chadwick, J. (1958). *The Decipherment of Linear B*. Harmsworth, Middlersex. UK: Penguin Books.

Chambers, N. C., Brakefield, J., Yahiel, D. I., & Fulgham, D. D. (1983). *Stress Assessment Through Voice Analysis.*, San Antonio Texas. Technology, Inc.

David, H. (1979). Development of an ATC Flow-Control Cell. *Ergonomics, 22*(5), 1(abstract).

David, H. (1985). *Measurement of Controllers' Mental State in a Real Time Simulation Environment* (183). EEC Report 183. EUROCONTROL Experimental Centre, Bretigny sur Orge, France,:

David, H. (1997). *Use of Microsaint for the Simulation of ATC Activities in Multi-Sector Evaluations.* Contemporary Ergonomics 1997

David, H. (2004). *The Effect of Human Intervention on Simulated Air Traffic.* Contemporary Ergonomics 2004,.

David, H. (2004). *Measures of Performance in Air Traffic Control.* Contemporary Ergonomics 2004,.

David, H. (2018). *The Air traffic Kludge,* Hastings,East Sussex, UK. R+D Hastings

David, H., Farbos, B., Bourgeois, S., Cabon, P., & Mollard, R. (1999). *Psychophysiological Measures of Adaptation to an Unfamiliar HMI in Simulated Air Traffic Control.* Contemporary Ergonomics 1999.

David, H., Mollard, R., Cabon, P., & Farbos, B. (2000). *Psychophysiological Measures of Adaptation to an Unfamiliar HMI in Real-Time ATC Simulation.* Contemporary Ergonomics 2000.

David, H., & Pledger, S. (1995). *Speech Recognition and Keyboard Input for Control and Stress Reporting in an Air Traffic Control Simulation.* Contemporary Ergonomics 1995.

Dubey, G. (2000). *Social Factors in Air Traffic Control Simulation.* EEC Report 348 EUROCONTROL Experimental Centre,Brétigny-sur-Orge, FRANCE:

Gawron, V. J. (2000). *Human Performance Measures Handbook.* New Jersey, USA.: Lawrence Earlbaum Associates,.

Holmqvist, K., Nystrom, M., Andersson, R., Dewhurst, R., Jarodzka, H., & Van der Weijer, O. (2011). *Eye Tracking, A comprehensive Guide to Methods and Measures.* Great Clarendon Street,Oxford, UK: Oxford University Press.

ISO 11064-3 (1999) *Control room layout* ISO 11064-3:1999/Cor 1:2002; Geneva, Switzerland: International Standards Organisation.

ISO 11064-1 (2000) *Principles for the design of control centres*; Geneva, Switzerland: International Standards Organisation.

ISO 11064-2 (2000) *Principles for the arrangement of control suites;* Geneva, Switzerland: International Standards Organisation.

ISO 11064-4 (2004) Layout and dimensions of workstations; Geneva, Switzerland: International Standards Organisation.

ISO 11064-6 (2005) *Environmental requirements for control centres;* Geneva, Switzerland: International Standards Organisation.

ISO 11064-7 (2006) *Principles for the evaluation of control centres*; Geneva, Switzerland: International Standards Organisation.

ISO 11064-5 (2008) *Displays and controls*; Geneva, Switzerland: International Standards Organisation.

Kendall, M. G. (1973). *Time-Series*. London U.K.. Charl;es Griffin and Company Limited.

Kendall, M. G. (1975). *Multivariate Analysis*London U.K. Charles Griffin & Company Limited..

Klimmer, F., Aulmann, H. M., & Rutenfranz, J. (1972). Urinary Catecholamine Elimination in Air Traffic Control Workers under Occupationally Induced Emotional and Mental Stress. *International Archives of Occupational Health,, 30*(1), 16.

Langford, J., & Deana, M. (2003). *Focus Groups*. London: Taylor and Francis.

Lee, A. T. (2005). *Flight Simulation - Virtual Environments in Aviation*. Farnham, Surrey, U.K.: Ashgate.

Mell, J. (1992). *Etude des Communications Verbales entre Pilote et Controlleur en Situation Standard et Non-Standard.* ENAC, Toulouse, FRANCE

Meshkati, N., Hancock, P. A., Rahimi, M., & Dawes, S. M. (1995). *Techniques In Mental Workload Assessment.* In J. R. Wilson & N. Corlett (Eds.), *Evaluation of Human Work 2nd.Ed.* London: Taylor and Francis.

Nightingale, F. (1858). *DIAGRAM of the CAUSES of MORTALITY in the ARMY in the EAST*. In Yau
(2013) *Data Points Visualization that Means Something*. Indianapolis,Indiana, USA:John Wilry and Sons Inc.

Oppenheim, A. N. (1992). *Questionnaire Design, Interviewing and Attitude Measurement*. London: Cassel :.

Pledger, S. (1994). *Effects of Self-Assessment on Performing Air Traffic Control.* Loughborough University, Loughborough Leicestershire LE11 3TU UK.

Rompelman, O., Van Kampen, W. H., Backer, E., & E, O. R. (1980). *Heart Rate Variability in Relation to Psychological Factors. Ergonomics, 23*, 15.

Rothschild, L. (1977). *Meditations of a Broomstick*. London: Collins.

SBFI (2015) *Control Centre Furniture – Solutions for every Sector*; London, UK: SBFI

Shadbolt, N., & Burton, M. (1995). *Knowledge Elicitation: A Systematic Approach* In J. R. Wilson & N. Corlett (Eds.), *Evaluation of Human Work 2nd. Ed*. London: Taylor and Francis. .

Siegel, S. (1956). *Nonparametric statistics for the behavioural Sciences*. New York, USA: McGraw-Hill Book Company.

Toga, A. W., & Mazziota, J. C. (1996). *Brain Mapping - The Methods*. California, USA: Academic Press, .

Wesson International, I. (1990). *The TRACON II Multi-player ATC Simulator,*. Dallas, Texas: Wesson International,.

Wilson, J. R., & Corlett, N. (1995). *Evaluation of Human Work 2nd.Edition*. London Taylor and Francis. .

Yau, N. (2011). *Visualise This The FlowingData Guide to Design, Visualisation and Statistics* Indianianapolis, Indiana, USA: Wiley Publishing, Inc.

Yau, N. (2013). *Data Points Visualization that Means Something*. Indianapolis, Indiana, USA: John Wiley and Sons, Inc.

Recommended Reading

Gawron, V. J. (2000). *Human Performance Measures Handbook*. New Jersey, USA. Lawrence Erlbaum Associates.

This is probably the widest-ranging survey of measures of performance that has been produced. However, it does not make any judgements of the value of the various methods, and it is in need of updating. (Lawrence Erlbaum Associates is no longer in business)

Holmqvist, K., Nystrom, M., Andersson, R., Dewhurst, R., Jarodzka, H., & Van der Weijer, O. (2011). *Eye Tracking, A comprehensive Guide to Methods and Measures*. Great Clarendon Street,Oxford, UK: Oxford University Press.

More than you will ever need to know about eye tracking.

Oppenheim, A. N. (1992). *Questionnaire Design, Interviewing and Attitude Measurement*. London: Cassel :.

A comprehensive guide to the design of questionnaires, with some good advice on interviewing and de-briefing. Does not cover modern on-line methods, such as Survey monkey (best googled).

Stanton, N.A., Salmon, P.A., Rafferty,L.A., Walker.G.H., Baber, C. and Jenkins,D.P. (2013) *Human Factors Methods, A practical Guide for Engineering and Design* Second Edition, pp 627:, Farnham, Surrey, U.K., Ashgate.

Wilson,J.R. and Corlett, E.N. (1995*) Evaluation of Human Work*. Second Edition, pp 1134: London, U.K. Taylor and Francis.

Either of these two monumental tomes introduces Human Factors or Ergonomics. Each is essentially a compilation of contributions from many experts, although Stanton et al (2013) is rather more integrated than Wilson (1995) is. One or other is essential reading if you are entering the field of Real-Time Simulation. If you are already in this field, they are more than essential, since you are probably out of date.

Lee, A. T. (2005*). Flight Simulation - Virtual Environments in Aviation*. Farnham, Surrey, U.K., Ashgate.

An excellent book on the use of specialised flight simulators, which reproduce the cockpits of aircraft. These simulators are usually very specific - and very expensive.

Yau, N. (2011). *Visualise This The FlowingData Guide to Design, Visualisation and Statistics* Indianianapolis, Indiana, USA: Wiley Publishing, Inc.

This is a particularly valuable resource, because Yau gives practical guidance on how to generate diagrams, tables and other images using both commercial and free software. He also gives valuable advice, with examples, on how to lay out these images. Unfortunately, the illustrations are often too small to see detail. To make proper use of this book, you need to spend several weeks downloading various computer packages, and virtually learning a complete computer language. Not for the faint hearted.

Appendix 1 - Measurement Methods

This appendix describes 33 types of measurement which have been applied in Real-Time simulation, divided (on the basis of bitter experience) into six useful general measurement methods, four possibly useful methods, ten individual measurement methods, four tentatively useful methods, and nine methods which are not recommended. It is usually necessary to use several types of measurement.

Aspects of Measurement Methods

Each method is classified on 16 different aspects that may be relevant in planning and analysing simulations. Where there is considerable variation, the codes given below may be accompanied by a question mark (**?**). Where the value is extreme, the code may be accompanied by an exclamation mark (**!**).

Each of the following tables is followed by brief notes on the individual methods.

1. Measurement type

    Measurements may be described as :

    i. **(EEG)** *Electroencephalography* which requires electrodes attached to the head, measuring microvolt differences.

    ii. **(EPhy)** *Electrophysiological* methods which usually require electrodes attached to the skin.

    iii. **(Obsn)** *Observation* which usually involves a human observer.

    iv. **(Physi)** *Physiological* measures which usually involve some medical apparatus.

    v. **(Subj)** *Subjective* measures which require the active participation of the operator.

    vi. **(Syst)** *System* measures which can usually be derived from the operating system.

2. Stress/Strain

    a) **(Ss)** *Stress* measurements measure the workload or "objective" difficulty of a task. They are usually specific to a particular work situation.

    b) **(Sn)** *Strain* measurements measure the effect of the workload on the operator. They are usually generally applicable over a range of working situations.

3. Time Scale

    The four levels of time scale correspond very roughly to traditional measurements. They give an idea of the speed of response of the measurement.

    a) **(D)** *Day* measurements refer to daily differences, such as a sleep log.
    These provide background measures, but rarely distinguish between alternative organisations.

    b) **(H)** *Hour* measurements relate to a period about the normal length of a simulation run.

    c) **(M)** *Minute* measurements give short-term measures that vary throughout the period of a simulation run. ISA measures are examples.

    d) **(S)** *Second* measurements relate to very rapid changes, such as eye-movement recording. These are usually too detailed to be useful in the analysis of a Real-Time simulation, although peak, 95% or mean values may be useful.

4. Portability

    Portability describes whether the method requires fixed equipment, or can be reasonably easily moved – for example from a simulator to a real operational control room.

    a) **(Y)** *Yes*. The method is portable.

    b) **(n)** *No*. The method is not easily portable.

5. Observer Effect

   The presence of observer effects suggests that measurements may be distorted because the controller is aware he is under observation. The direction of the distortion is not always obvious, and its quantitative effect is rarely measurable.

   a) **(Y)** *Yes.* The method is subject to observer effects.

   b) **(n)** *No.* The method is not subject to observer effects.

6. Failure Risk

   Failure risk is the risk that the measurement may not actually work. For example, pupil dilation, sometimes used as a measure of arousal, is also affected by changes in ambient lighting, or even by the illumination level in the direction the eye is looking – such as from a dark working surface to a luminous screen.

   a) **(Y)** *Yes.* The method is subject to failure risk.

   b) **(n)** *No.* The method is not subject to failure risk.

7. Bias Risk

   Bias risk is the risk that the method may be influenced by something unexpected (for example salivary cortisol is strongly affected by the time of day.) Naturally, it is not always possible to expect the unexpected, but experience has shown that some methods are particularly likely to fail in this way.

   a) **(Y)** *Yes.* The method is often subject to bias risk.

   b) **(n)** *No.* The method is not usually subject to bias risk.

8. Ethical Problems

Ethical problems may occur where the use of certain methods is being considered. It is always wise to consider the possibility of an ethical problem before using the method. The occurrence of such a problem during a simulation may require on-the-spot decisions with potentially disastrous consequences.

   a) *(+) Medical.* The method is subject to problems of medical ethics. For example, the experimenter may discover that an observer has a colour vision defect that makes it dangerous for him to perform the task, or has a significantly abnormal heart rate. **The diagnosis of medical conditions may only be carried out by qualified medical personnel.** Unqualified experimenters must abstain from any such activity.

   b) **(P)** *Privacy.* The method may produce information that ought not to be made public without the express permission of the operator. Special care must be taken when handling such data.

   c) **(n)** *No.* The method is not usually subject to ethical problems.

9. Setup Cost.

Set-up cost usually means equipment plus software. Some measures have alternative methods, which vary greatly in cost. Some methods use generally available software, much of which is available in 'open-source' form.

   a) **(H)** *High.* The method requires considerable cost to setup For example, modern eye-movement systems will cost tens of thousands of dollars, euros or pounds.

   b) **(M)** *Medium.* The method requires some investment, particularly if initial staff training is required. Activity Analysis, particularly where one observer measures several operators, requires careful training.

   c) **(L)** *Low.* The method requires little or no investment. For example many system measures are available as a matter of routine, and do not need any additional treatment.

10. Running Cost.

    Running cost covers expendable stores. Most methods have relatively low expendable stores cost.. Some measures have alternative methods, which vary greatly in cost. Where there is considerable variation, the code given below may be accompanied by a question mark (**?**).

    a) **(H)** *High.* The method requires considerable cost to run. For example, analysis of body fluids is always expensive.

    b) **(M)** *Medium.* The method requires some investment.

    c) **(L)** *Low.* The method requires little or no investment. For example many system measures are available as a matter of routine, and do not imply any additional running cost.

11. Staff needed.

    Staff cost may depend on the context. In many academic situations, undergraduate or postgraduate students are virtually slave labour, which is definitely not the case in industry. Some measures have alternative methods, which vary greatly in cost

    a) **(H)** *High.* The method requires one or more observer per operator. For example, EEG methods require at least one or sometimes two highly trained observers for each operator measured.

    b) **(M)** *Medium.* The method requires one observer with no special skills.

    c) **(L)** *Low.* The method requires no additional staff.

12. Analysis cost.

The cost of analysis depends very much on the nature of the data. Where human judgement is required, or where specialised proprietary software is required, analysis can be very costly. To an increasing extent, data are recorded in digital form, which makes analysis relatively cheap.

   a) **(H)** *High.* The method requires detailed examination of records, and human judgement in coding the data. For example, traditional EEG records can be analysed by Fourier Analysis, but a trained, experienced observer is required to interpret the results of the analysis.

   b) **(M)** *Medium.* The method requires one observer with no special skills. For example, the transcription of de-briefing notes.

   c) **(L)** *Low.* The method requires no additional staff.

13. Analysis speed.

The speed of analysis depends very much on the nature of the data. Where human judgement is required, analysis can be very slow. Equally, where specialised laboratory techniques are required, as in the assessment of catecholamine or melatonin, samples may need to be transported to a remote laboratory. To an increasing extent, data are recorded in digital form, which makes analysis fast, if proper analytic software is available.

   a) **(H)** *High.* The method can produce results directly, or with very little processing. For example, ISA performed with modern software.

   b) **(M)** *Medium.* The method some effort, but produces results within a week or so.

   c) **(L)** *Low.* The method requires external effort, such as catecholamine or other hormonal levels. Samples may have to be frozen, transported to a hospital or commercial analytic laboratory and wait in a queue with other clients, before results become available.

14. Automatic Data Collection

The cost of data collections depends very much on the nature of the data. Where human judgement is required in recording data, collection can be very costly. To an increasing extent, data are recorded in digital form, using the simulation software, which makes data collection very cheap.

a) *(Y) Yes.* The data for this method can be collected automatically.

b) *(n) No.* The data for this method cannot be collected automatically.

15. Automatic Data Analysis

The cost of analysis also depends very much on the nature of the data. Where standard analysis techniques can be used, automatic analysis is relatively cheap.

a) *(Y) Yes.* The data for this method can be analysed automatically.

b) *(n) No.* The data for this method cannot be analysed automatically

16. Utility of Method

This is an evaluation of the best course of action when this method is proposed. It is based on extensive practical experience. Sadly, there is very little progress in this field, although some methods are promising. Unfortunately, at different times, different methods are considered promising, producing a misleading impression of progress.

a) *Regular Use.* These methods have been in regular use, and are generally recommended

b) *If Possible.* These methods are used where they can be applied. This is not always the case.

c) *Available.* These methods are available for use, but experience suggests that they are not worth the effort except in special cases.

d) *Individual.* These methods can be applied to one or two operators, but are not suitable for mass application, usually because they require extensive setting-up and monitoring.

e) *Tentative.* These methods have potential value, but have not, so far, proved reliable in Real-Time simulations.

f) *Negative.* These methods are not recommended.

Measures Recommended for regular use

This table summarizes six methods recommended for regular use.

Measurement	Type	Stress/Strain	Time scale	Portability	Obs. effect	Failure risk	Bias risk	Ethical probs.	Setup cost	Running cost	Staff needed	Analysis cost	Anal. speed	Auto. data	Auto. analysis	Utility of the method
Op. Records	Syst	Ss	M	n	n	n	n	n	L	L	L	L	H	Y	Y	Regular Use
NASA-TLX	Subj.	Sn	H	Y	n	L	L	n	L	L	L	L	H	Y	Y	Regular Use
On-line ISA	Subj.	Sn	M	n	n	n	L	n	H	L	L	L	H	Y	Y	Regular Use
Debriefing	Subj.	Sn	H	Y	Y	M	M	n	L	H	L	M	H	n	n	Regular Use
Questionnaires	Subj.	Sn	D	Y	n?	M	M	n	L	L	L	L	S	Y	Y	Regular Use
Embedded Tasks	Syst	Sn	M	n	n	L	L	n	H	L	L	L	H	Y	Y	If Possible

Symbols

Ss Stress	(D) Days	(Y)es	(H)igh	? = Maybe	! = Very
Sn Strain	(H)ours	(n)o	(M)edium	+ = Medical	P = Privacy
	(M)inutes		(L)ow		

Subj = Subjective
Syst = System

Useful General Measurement Methods

Operational records are the records produced by the simulator. In ATC, for example, they would include the entry and exit times and points for each simulated aircraft for each sector, the start and end times of communications links, the times and contents of written messages, and many other items. In many simulators, false starts and cancelled messages are not normally recorded, although the nature and frequency of these may be very important in assessing the efficiency of an interface.

The **NASA-TLX** is a Task Load Index developed by NASA, and widely used as a measure of the subjective difficulty of a task. It involves six different scales, representing different aspects of the task. It is fully described in Chapter 10 and an example of the form used is given in Chapter 11. This measure is recommended to be taken as soon as possible after the end of a simulation.

The On-Line ISA (Instantaneous Self-Assessment) is a measure using a scale of 1-5 to express levels of activity from 1 (Very Low) to 5 (Very High). It is fully described in Chapter 10.. This measure requires prepared software, and should be applied at regular intervals – three minutes is usual - throughout the simulation.

Debriefing is the practice of discussing the simulation immediately after completion with the operators concerned. It is a selective process, requiring judgement on the part of the debriefer. The loudest voices are not necessarily the most important, and skill is required to draws out the feelings or opinions of individuals. The reactions of operators will depend very strongly on cultural factors. Some cultures are reluctant to speak out in the presence of older colleagues, and some are reluctant to speak at all. Debriefing is a two-way process, and both objectivity and empathy are required in an effective debriefer. Some hints can be gathered from Langford and McDonagh (2003), although this book is primarily concerned with Focus Groups, where the participants are not usually expert in the field..

Questionnaires are widely used to provide structured responses to specific questions, and to bring out complex responses. Considerable skill and experience are needed to construct questionnaires, and strict precautions must be taken to maintain confidentiality, and to convince the operators that their privacy will be respected. Ideally, Individual responses should not be reported. If it is necessary to do so, great care should be taken to ensure anonymity.(Oppenheim, 1992)

Embedded Tasks can be useful if they are available. These are tasks that are not strictly necessary, or can be performed by automated back-up systems. Under pressure, the operator may simply omit them. The level of difficulty of a task may be judged by the extent to which operators stop carrying out these tasks. Unfortunately, it requires detailed acquaintance with the system (and the operators' attitude to these tasks) to identify embedded tasks, and, in many systems, they simply do not exist.

Possible Measures for regular use

This table summarizes four methods that may possibly be used regularly.

Measurement	Type	Stress/Strain	Time scale	Portability	Obs. effect	Failure risk	Bias risk	Ethical probs.	Setup cost	Running cost	Staff needed	Analysis cost	Anal. speed	Auto. data	Auto. analysis	Utility of the method
Sleep logs	Subj.	Sn	D	Y	n	L	H	n	L	L	L	L	M	Y	Y	Available
Behaviour	Obsn.	Sn	M	Y	Y	Y	Y	n	L	L	L	L	H	n	Y?	Available
Activity Analysis	Obsn.	Ss	M	Y	Y	L	M	n	L	L	L	L	M	Y	Y	Available
Posture Analysis	Obsn.	Ss	M	Y	Y	M	M	n	L	L	L	L	M	n	Y	Available

Symbols

Ss Stress	(D) Days	(Y)es	(H)igh	? = Maybe	! = Very
Sn Strain	(H)ours	(n)o	(M)edium	+ = Medical	P = Privacy
	(M)inutes		(L)ow		

Obsn = Observation
Subj = Subjective

Sleep Logs are a precautionary method, used mainly during extended simulation or other trials. It sometimes becomes clear that participants are accumulating fatigue, biasing their subjective assessments of the work they are doing. In addition to ensuring that the workloads are reasonable, it is a wise precaution to monitor the fatigue and sleepiness of the participants.

Each simulation participant is issued with a booklet, in which each page corresponds to a day of the simulation. The operator records each day the time he went to bed, the time (approximately) that he fell asleep, any interruptions, woke, and got up, his states of fatigue and sleepiness on going to bed and rising. If possible, the use of the sleep log should begin some days before the experiment and continue for some days after its end. It is, however, not easy to ensure that logs will be returned after the operators have left the simulation centre, and it may be advisable to collect logs before participants leave – giving them a second log for the following week to be returned in a return addressed, sealable, stamped envelope.

Behaviour is obviously a potential measure of strain on operators. The difficulty lies in the definition of what behaviour is. Until recently most behavioural observation was essentially anecdotal – lacking in any real validation. Even now, there are considerable cultural and professional differences in the behaviour expected of operators. Air Traffic Controllers and Air Crew pride themselves on calm and unruffled responses to emergencies, to the extent that emergencies have sometimes not been recognised.

Activity Analysis is a simple, empirical technique. A finite number of observable activities is defined during initial observation of activities in the workplace, in consultation with experts in the field of activity. This is in reality not as simple as it may appear. (Many of the problems of 'knowledge elicitation' discussed in Shadbolt and Burton (1995) apply even at this level. In particular, many expert operators are what they describe as 'Samurai' – who perform a task in an 'automated' manner without being able to describe it in words.) One observer can record the activities of a complete control room on a sampling basis, or two or three operators continuously.

Posture Analysis is similar to Activity Analysis, except that the observer records the physical position of the operator. It is a measure of stress, but most useful where heart rate or other physiological measures are being made, to try to separate the usually dominant effects of physical activity, such as standing up, or raising the arms when seated.

Individual Measurement Methods

This table summarises ten measurement methods, which may be useful, but which can only be practically applied to one or two individuals in Real-Time Simulation.

Measurement	Type	Stress/Strain	Time scale	Portability	Obs. effect	Failure risk	Bias risk	Ethical probs.	Setup cost	Running cost	Staff needed	Analysis cost	Anal. speed	Auto. data	Auto. analysis	Utility of the method
Heart Rate	Physi	Sn	M	n	n	M	Y	+	H	H	H	L	M	Y	Y	Individual
Sinusarrythmia	Ephy	Sn	M	n	n	M	Y	+	H	H	H	L	M	Y	Y	Individual
Blood Pressure	Physi	Sn	M	n	n	L	n	+	H	H	H	H	M	Y	Y	Individual
Alpha Rhythm	EEG.	Sn	M	n	n	n	n	+	H	L	L	Y	L	Y	Y	Individual
Beta Rhythm	EEG.	Sn	M	n	n	n	n	+	H	L	L	Y	L	Y	Y	Individual
Delta Rhythm	EEG.	Sn	M	n	n	n	n	+	H	L	L	Y	L	Y	Y	Individual
Theta Rhythm	EEG.	Sn	M	n	n	n	n	+	H	L	L	Y	L	Y	Y	Individual
Ev.Cerebral Potl.	EEG.	Sn	M	n	n	n	n	+	H	L	L	Y	L	Y	Y	Individual
Respiration	Physi	Sn	M	n	n	M	n	+	H	H	H	M	M	Y	Y	Individual
Eye Movement	Ephy	Ss	S	n	n	L	L	n	H	H	H	H!	L	Y	Y	Individual

Symbols
Ss Stress   (M)inutes   (Y)es   (H)igh   ? = Maybe   ! = Very
Sn Strain   (S)econds   (n)o   (M)edium   + = Medical   P = Privacy
                                      (L)ow

Physi = Physiological
Ephy = Electrophysiological
EEG. = Electroencephalography

Heart Rate is the most widely used measure of physiological strain. Many cheap devices are available for measuring heart rate, some of which can be operated remotely. Most cheap devices are not very reliable, and do not store a continuous record. Heart rate is subject to considerable variation, and may have medical implications, which should be avoided by non-medical researchers. Since heart rate is affected by physical effort, some form of posture analysis should be employed as a control.

Sinusarrythmia is the irregularity of heart rate. In medical practice, sinusarrythmia refers to the severe disturbances of the regular beating of the heart, and indicates imminent heart failure. In this context, we are concerned with much smaller variations. Measurement requires recording of beat-to-beat intervals. Heart rate slows down as the chest expands, and a spectral analysis will usually show a component corresponding to respiration rate. Unfortunately, speech upsets the regular rhythm of breathing, and most operators use speech as part of their control activity. To control for chest expansion, a continuous measure of chest expansion is also required. Generally, it is not worth the effort.

Curiously, mental effort is associated with a reduction in sinusarrythmia, not an increase.

Blood Pressure is clearly associated with strain. Unfortunately, measurement of blood pressure usually requires interference with the operator's primary task. Posture, anticipation and many other factors affect blood pressure. (One excursion in blood pressure was traced to the operator dropping a lighted cigarette end inside his shirt - some years ago.) At present, it is probably better avoided.(Rompelman, Van Kampen, Backer, & E, 1980)

Electroencephalography in general has developed empirically, using the available tools. Since it is now clear that the brain does not use continuous electrical signals, this is like trying to work out what is happening in a computer by measuring the electromagnetic field around it. Studies using conventional classical correlation produce frustrating hints, but do not lead anywhere

Delta, Theta, Alpha, Beta Rhythms are rhythms derived by Fourier analysis, a mathematical technique for identifying regular rhythms in apparently irregular signals. The irregular signals in this case are picked up by electrodes attached to the scalp. A period of several seconds is necessary to provide sufficient data for analysis. The characteristic frequency bands, in order are:

Delta (1-3 Hertz) - associated with certain types of brain defects and also with 'strain' in general terms. The association is not well defined.

Theta ((4-7 Hertz) – associated with learning. Fairly well correlated.

Alpha (8 – 12 Hertz) – associated with 'not thinking' – fairly well defined and associated with 'meditation' or 'idling'. Probably the only useful measure in practical contexts.

Beta (16-30 Hertz) – associated with mental work. The higher range appears to be associated with visual work – trained observers could identify when a controller was using a PPI (radar-like) display, rather than reading and calculating height information.

Evoked Cortical Potential is a method of assessing the mental load on an operator by asking him to count rare signals in a stream of sound signals presented during a task. It appears to work, but the mechanism is not clear, and the method may interfere with the operator's real work. It does suggest that any measurable 'mental fatigue' effects are transient (a matter of seconds for full recovery).

More modern, very expensive and intrusive techniques (Toga & Mazziota, 1996) have begun to discover what parts of the brain carry out specific activities. However, these activities do not correspond well to traditional concepts of thought, and it is clear that we do not yet understand brain function at any practically useful level.

In principle, it should be possible using surface electrodes, to identify which parts of the brain are active. In practice, there is a long way to go before these measures are practically useful.

Respiration can be measured with a chest-band and a tension sensor, or via EKG sensors, or a thermistor in the nostril (Casali & Wierwille, 1984). Although not important in itself, the effects of respiration on heart rate mean it has to be measured as an intervening factor. In clinical studies, it can be compensated for by identifying a component after Fourier analysis at around 0.1 Hertz. This requires that the operator does not speak. If the operator has to speak as part of his task, then an empirical model, taking into account the actual expansion of the chest, must be constructed for each operator. This is generally not worthwhile.

Eye Movement studies have improved considerably in recent years. It is now possible to obtain a spectacle-type device that indicates on a video-recording where the operator is looking, although it is, so to speak. eye-wateringly expensive. The interpretation of eye-movement data is more complex. It is not yet possible to integrate eye-movement data with input or display details. Eye movement records are more useful to demonstrate that the operator does not look at a particular display or display window. Equally, it cannot be assumed that a display is better because the operator spends less time looking at it. Operators have to look somewhere, and looking does not always mean seeing.

If it is not practical to analyse the direction of gaze, it may be useful to measure the number and direction of fixations. Increased frequency and reduced duration may indicate strain on the operator. These statistics should be part of the software package of any eye movement device.

Blink rate can also be determined from eye-movement records – see below.

Holmqvist et al. (2011) is a comprehensive guide to the current state of the art of eye-tracking. Unfortunately, equipment capable of providing reliable records is very expensive, and can need considerable set-up time.

Even where adequate recordings can be obtained, there remains, as with electroencephalography, the very difficult problem of determining what the recordings mean. As a controller pointed out to me, a longer time spent looking at a display does not necessarily mean it is more difficult to use – it may simply be more interesting, particularly if it is a novel design.

Conversely, however, the observation that the controller does not look at a display may suggest that it is redundant. Even here, caution is needed. In one ATC simulation, the controllers appeared to glance at the strips presented only once, at the start. It transpired that they were looking at the code for the destination airport. They rarely looked back at it, until, inevitably, they attempted to land an aircraft at the wrong airport. Having done this once, they tended to check regularly before starting the approach.

Tentative Measurement Methods

This table summarizes four tentative measurement methods. These methods have been used, but require more study and evaluation before they can be recommended.

	Measurement Type	Stress/Strain	Time scale	Portability	Observer effect	Failure risk	Bias risk	Ethical problems	Setup cost	Running cost	Staff needed	Analysis cost	Analysis speed	Automatic data	Auto. analysis	Utility of the method
Ambient Sound	Obsn.	Ss	M	Y	n	Y	M	n	M	L	L	L	M	Y	Y?	Tentative
Blink Rate	Physi	Sn	H	n	n	Y	Y	+	H	H	H	H	M	Y?	Y?	Tentative
Body Fluids	Physi	Sn	D	n	n	M	H	+	H	H	H	H	L	n	Y	Tentative
Voice Stress	Syst	Sn	M	n	n	H	L	n	H	L	L	L	S	Y	Y	Tentative

Symbols

Ss Stress	(D)ays	(Y)es	(H)igh	? = Maybe ! = Very
Sn Strain	(H)ours	(n)o	(M)edium	+ = Medical
	(M)inutes		(L)ow	

Obsn = Observation
Physi = Physiological
Syst = System

Ambient Sound is an opportunistic measure of general stress. It depends on the observation that a normal control room is relatively quiet, and that increased activity is usually associated with increased noise. Unlike other measures, it applies to the entire control team. Its main virtue is its cheapness – 'never mind the quality, feel the cost'. A standard recording noise meter can be used, which should produce a digital file of noise intensity at regular intervals.

Blink Rate can be obtained as a by-product of eye-movement recording, from EEG recording, where blinking may disturb recording, and an additional electrode is used to detect it, or from manual analysis of video-records. It is liable to distortion by environmental factors, which may affect the drying of the cornea. In principle, this should not affect the use of blinking to assess differences between similar situations. There is a body of studies that suggest an association between blinking and thought, as far as thought can be defined. Slightly different experimental layouts might make a considerable difference to blink rates if sharp differences in luminance are involved.

Body Fluids cover a considerable range of biological chemicals, of which Cortisol and Melatonin are examples. Although they are undoubtedly related to the strain on the operator, they are usually slow, costly and intrusive. Analysis of saliva is probably the least intrusive. Almost all methods require analysis in clinical laboratories, requiring the preservation, transportation and storage of samples. Some of the methods used by anti-doping agencies in commercial 'sport' may be useful in this respect, but both accidental and deliberate contamination or deterioration of samples is possible . Analysis is, in any case, a matter of weeks or days rather than hours. Almost all measures of body fluids are subject to known and unknown sources of bias, to the extent that even the most reliable measures may be distorted. Rutenfranz (1972) describes a study where catecholamine appeared to increase in relaxed working, in contradiction to accepted theory. Further enquiry discovered that the controllers drank coffee when not working, and that caffeine by-products were mistaken for catecholamine in analyses. Fundamentally, we do not really know body metabolism in general, and its variations between groups, ethnic or social, sufficiently well for these measures to have predictive or even explanatory value.

Voice Stress analysis is based on the observation that the fundamental frequency of speech rises under conditions of stress. Sections of speech about 20 seconds long are required, and a Fourier analysis, similar to that used in EEG work is used to identify the fundamental frequency. This has to be compared to a 'resting' state for each operator. Air traffic controllers pride themselves on their calm speech, and may not show significant variations. Interest in the method has flagged in recent years, suggesting that it may not be practical (Chambers, Brakefield, Yahiel, & Fulgham, 1983).

Non-recommended Measurement Methods

This table summarizes nine non-recommended measurement methods. These have been tried and found not to work in practice.

Measurement Type	Stress/Strain	Time scale	Portability	Observer effect	Failure risk	Bias risk	Ethical problems	Setup cost	Running cost	Staff needed	Analysis cost	Analysis speed	Automatic data	Auto. analysis	Utility of the method
Muscle Tension	Physi	Sn	M	n	?	M	Y	+	H	H	H	H	S	Y?	Y?	Negative
Skin Resistance	Ephy	Sn	M	n	L	H	n	+	H	H	H	H	S	Y?	Y?	Negative
Tremor	Physi	Sn	M	n	n	H	L	+	H	H	H	H	S	Y?	Y?	Negative
SWAT On-line	Obsn.	Sn	M	n	n	n	L	n	H	L	L	L	H	Y	Y	Negative
Struct. Interview	Obsn.	Sn	D	Y	M	L	H	n	L	H	H	H	S	n	n	Negative
Secondary Tasks	Obsn.	Sn	S	n	n	H	H	n	H	L	L	H	M	Y	Y	Negative
KL Test	Subj.	Sn	H	Y	n	n	H	n	L	L	L	L	M	Y?	Y?	Negative
GR Test	Subj.	Sn	H	Y	n	n	H	n	L	L	L	L	M	Y?	Y?	Negative
Peer Asssssment	Obsn.	Sn	D	Y	Y	M	M	P	L	L	H	L	L	n	n	Negative

Symbols

Ss Stress	(D) Days	(Y)es	(H)igh	? = Maybe	! = Very
Sn Strain	(H)ours	(n)o	(M)edium	+ = Medical	P = Privacy
	(M)inutes		(L)ow		
	(S)econds				

Obsn = Observation
Physi = Physiological
EPhy = Electrophysiological
Subj = Subjective

Muscle tension is an electrophysiological measure, which measures the state of the operators' physical processes, without regard to the 'objective' workload, and in consequence measure 'Strain'. Most control operations involve some form of physical activity, which may be proportional to workload (stress). It is therefore necessary to find muscles that are not involved in physical work related to control activity.

Muscle tension appears in response to stress, but strain-related tension may appear before the actual events defining the stress, particularly where the operator has encountered similar situations before. Time scales empirically appear to be of the order of minutes. Electromyography (EMG) measures usually require attaching the operator to a trolley of recording equipment. Although R/T or Infra-red free-movement systems are available, they may experience transmission problems in simulator or control room environments. Electrophysiological methods in general are not usually subject to observer effects. However, any system involving skin electrodes may fail if the electrodes become detached. In addition, different individuals are more or less affected physiologically by 'mental' strain, and some show no measurable variations in muscle tension. Muscle tension may be affected by physical movements of the operator. In principle, a muscle should be selected that has no part in normal physical activity related to the task. In practice, it is difficult to find one, since the human muscular system does not operate like a computer-controlled system of levers, containing some apparently illogical responses. So far, there is no sufficient body of work to justify using this method.

Skin Resistance varies in response to mental strain. (This is the essential justification of the so-called 'lie-detector'). Skin resistance can be measured using surface electrodes. It requires the application of a small DC voltage to measure the resistance, which has on occasion led to blistering of the skin and the abandonment of the method. Skin resistance reacts rapidly to stimuli, usually by a sudden change, which then returns to a base level. In order to obtain a consistent response, the data must be integrated over some time. Some operators can control their skin resistance by conscious or unconscious means.

Tremor, particularly in wrist and finger muscles varies in response to mental strain. It can be measured by an accelerometer-like device mounted in a finger-ring, which is worn by the operator. Some individuals are more or less affected physiologically by 'mental' strain, and some show no measurable variations in tremor. Some individuals show a permanent state of 'tremor' (which is reported to be hereditary in part.) Tremor may be induced by a variety of routine medicaments and more or less legal substances. It may also be a symptom of serious illness, and should not be used by observers without medical qualifications.

SWAT, the Subjective Workload Assessment Technique, is a self-reporting technique. The operator is asked to report how stressed (s)he is on three scales – Time Load, Mental Effort Load and Psychological Stress load. Each scale has three levels. It is fully described in Chapter 10. SWAT requires a ranking of the different scales by a complex method, requiring a computer-based weighting program. Unfortunately, the designated supplier of this software appears to be defunct – rendering the method generally unavailable. (In any case, a comparative study suggests that it is significantly more intrusive than ISA and does not produce independent scales in use.(David & Pledger, 1995, Pledger, 1994,)

Structured interviews require a trained interviewer to ask an operator for his opinions and suggestions on a series of specific aspects of the exercise, according to a prepared script of questions. Interviewers are trained to draw out the operator, and may follow up unplanned aspects if they seem important. It is a major skill of the interviewer to extract an impression of what the operators really feel about the organisations being examined. (Structured Interviews can only be held after enough exercises have been completed for the operators to have formed opinions.)

In principle, structured Interviews, since they need no equipment, could be held wherever and whenever convenient - after a working shift in an operational centre, for example. Operators, like any other interviewees, are extremely sensitive to minor, nearly subliminal signals from the interviewer, which may lead to operator effects. There is a moderate risk that the interviewer may not be able to obtain definite responses to some questions. Inexperienced interviewers or those unfamiliar with the culture, language or attitudes of the operators may lose their confidence, and will obtain no useful data. Operators generally do their best to give truthful, realistic answers to the problems they see. There is, however, a moderate risk that they may give responses that are either what they think the group of which they are part will think, or what they think the observer wants to hear.(Langford & Deana, 2003),(Oppenheim, 1992)

Secondary tasks are additional tasks, not forming part of the primary task, which the operators are asked to perform when not fully occupied by the primary task. Ideally, the amount of effort that the controller makes available for the secondary task represents 'spare mental capacity'. Secondary tasks have been used in laboratory experiments on the assumption that subjects have a certain fixed capacity for work, and can distribute this capacity between a primary and a secondary task. The primary task, in this instance would be Air Traffic Control, and the secondary task some form of mental task.

Meshkati, Hancock and Rahimi (1995), in Wilson and Corlett (1995) discuss secondary task methods in some detail. It appears clear that it is not realistic to assume that controllers could move effortlessly from a primary ATC task to a secondary task and back, maintaining full efficiency in the ATC task. Indeed, On-line ISA (QV) or SWAT (QV), which can be considered to be simple secondary tasks, can be shown to lead to deterioration in the efficiency of the ATC task (David and Pledger 1995). It appears to be very difficult to strike a balance between a secondary task, which, to be measurable, must involve some explicit activity, and a primary task which involves a substantial amount of monitoring. In aviation studies, the effect of loss of attention shows up rapidly in deviations from the planned flight path, but failures of ATC monitoring and planning may not have consequences for many minutes, if at all.

Embedded secondary tasks (QV), however, are tasks that form part of a complex primary task, but which can be omitted without serious consequences for the overall efficiency of the primary task. Their characteristics are sufficiently different from those of classical secondary tasks to merit separate discussion.

The **Konzentrations-Leistung** test is a test of the ability to carry out mental operations involving mental arithmetic, retaining one result and comparing with a second. In principle, this is a test of mental capacity, used to identify strain. This test can be applied before and after an exercise, with the intention of measuring a possible decrease in mental efficiency during the exercise. Because it is a before-and-after test, which is applied at the start and end of an exercise, it can only measure effects over the entire exercise. This test aims at identifying a decrement in logical reasoning ability following heavy mental work. Unfortunately, the inevitable time lag between the end of the Real-Time simulation run and the start of the test appears to allow time for the controller to recover from any work-related effects. (David, 1985)

This test is not generally available for research work, since it is a copyrighted test sometimes used in psychiatry. It cannot therefore be recommended for general use.

The **Grammatical Reasoning test** is a test of the ability to carry out mental operations involving chains of logical reasoning (Baddeley, 1988; Rothschild, 1977). The operator is presented with a test sheet containing statements of the form:

"A/B is/is not preceded/ followed by B/A "   - "AB/BA".   True?: False?

(Where the slashes separate alternatives) and is required to circle the correct statement (True?: False?)

In principle, this is a test of mental capacity, used to identify strain. This test is applied before and after an exercise, with the intention of measuring a possible decrease in mental efficiency during the exercise. Because it is a before-and-after test, which is applied at the start and end of an exercise, it can only detect differences over the entire exercise. This test aims at identifying a decrement in logical reasoning ability following heavy mental work. Unfortunately, the inevitable time lag between the end of a Real-Time simulation run and the start of the test appears to allow time for the controller to recover from any work-related effects.

A specific cross-cultural bias appears to occur with this test. Controllers whose native language is German appear to find the passive negative construction particularly difficult. (Some controllers could not answer the question: 'B is not preceded by A -AB') All controllers preferred to take this test in English. Some individuals who could not answer questions in German could do so in English, although this was not their native language. This raises some interesting questions about mental capacity as such – whether one individual's mental capacity can vary when using a different language, or whether the structure of a language or the way it is normally used can affect reasoning capacity.

Appendix 2 – Report

The following pages contain the complete report of the dummy simulation used as an example in this book.

Any resemblance to other simulation reports may be coincidental.

"If you can't take a joke, you shouldn't have joined!"

(Traditional UK Service comment on any disaster, from a broken fingernail to a broken leg.)

Hugh David

Control Room Simulation

EREHWON ORGANISATION

FOR THE SAFETY OF AIR NAVIGATION

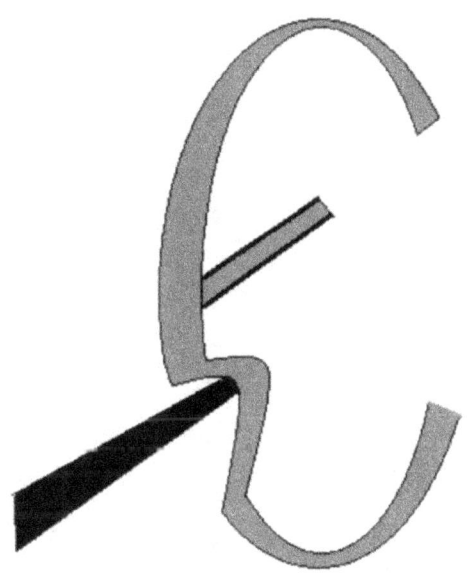

Hugh David

Control Room Simulation

EREHWONTROL EXPERIMENTAL CENTRE

Ruritania I

Evaluation of a two sector division of Ruritanian Airspace

and a Semi-automatic Sequencing System

DUMMKOPF

for Strelsau International Airport.

EEC Report No. 666

Issued: March 2020

The information contained in this document is the property of the EREHWONTROL Agency and no part should be reproduced in any form without the Agency's permission.

The views expressed herein do not necessarily reflect the official views or policy of the Agency.

Intentionally Left Blank

(Heaven knows why)

Control Room Simulation

REPORT DOCUMENTATION PAGE

Reference:	Security Classification:
EEC Report No. 666 (Ruritania I)	Unclassified
Originator: EEC – SIM (SIMulation Division)	**Originator (Corporate Author) Name/Location:** EREHWONTROL Experimental Centre Utopia City Utopia
Sponsor: R. Hentzau	**Sponsor (Contract Authority) Name/Location:** Ministry of Aviation Air traffic Control Authority Strelsau International Airport Strelsau, Ruritania

TITLE:

RURITANIA I

Evaluation of a two sector division of Ruritanian Airspace

and a Semi-automatic Sequencing System for Strelsau International Airport.

Author	Period	Pages	Figures	Tables	Attachment	References
E. Bennet EREHWONTROL Experimental Centre	October 2020	VI + 65	54	-	=	-

Distribution Statement:

(a) Controlled by: E. Bennet
(b) Special Limitations: None
(c) Copy to NTIS: YES (*God knows what they will do with it!*)

Descriptors (keywords):

Air Traffic Control, Ruritania, Sectorisation, DUMMKOPF, Sequencing

Abstract:

A Real-Time Simulation of Ruritanian Airspace has shown that a division into two sectors (Ruritania East and Ruritania West) will lead to an increase of about 25% in weekday and 45% in Weekend traffic capacity.

The DUMMKOPF system, however, is not yet in a suitable state for implementation.

This document has been collated by mechanical means. Should there be missing pages, please report to:

> EREHWONTROL Experimental Centre
>
> Publications Office
>
> Utopia City
>
> Utopia

Control Room Simulation

EEC Report No.666

Issued: March 2020

RURITANIA I

Evaluation of a two sector division of Ruritanian Airspace

and a Semi-automatic Sequencing System for Strelsau International Airport.

By

E. Bennett

EREHWONTROL Experimental Centre

SUMMARY

A Real-Time Simulation of Ruritanian Airspace has shown that a division into two sectors (Ruritania East and Ruritania West) will lead to an increase of about 25% in Weekday and 45% in Weekend traffic capacity.

The DUMMKOPF system, however, is not yet in a suitable state for implementation

Hugh David

A2 - vii

Control Room Simulation

Contents

Contents...A2-1

Figures..A2-2

Abbreviations..A2-4

Executive Summary..A2-5

Chapter 1 - Introduction..A2-6

Chapter 2 - Questions Raised..A2-8

Chapter 3 - Ruritanian Airspace...A2-9

Chapter 4 - Experimental Plan and Measurements......................A2-16

Chapter 5 - Conduct of the Simulation..A2-25

Chapter 6 – Analysis...A2-26

Chapter 7 – Conclusions...A2-64

Acknowledgements...A2-65

Figures

Figure 1 - Ruritania Airspace..A2-9

Figure 2 - Overall Traffic 95% Weekday...A2-10

Figure 3 - Strelsau International 95% Weekday Traffic..............................A2-10

Figure 4 - Hentzau 95% Weekday Traffic...A2-11

Figure 5 - Zenda 95% Weekday Traffic A2-11

Figure 6 - Overall Traffic 95% Weekend..A2-12

Figure 7 - Strelsau International 95% Weekend Traffic.............................A2-12

Figure 8 - Hentzau 95% Weekend Traffic..A2-13

Figure 9 - Zenda 95% Weekend Traffic...A2-13

Figure 10 - Ruritania Airspace – 2 Sector Initial Proposal.........................A2-14

Figure 11 - Ruritania Airspace – 2 Sector Revised Proposal.....................A2-14

Figure 12 - Coding of Simulation Conditions..A2-16

Figure 13 - Simulation Experimental Plan..A2-17

Figure 14 - ISA Values..A2-18

Figure 15 - NASA-TLX Scales...A2-19

Figure 16 - Massie Grid - Sectorisation...A2-23

Figure 17 - Massie Grid - TMA..A2-24

Figure 18 - Sectorisation Anovar Factors..A2-27

Figure 19 - Sectorisation ANOVAR Sector Level.....................................A2-28

Figure 20 - Sector Level Weekday..A2-29

Figure 21 - Sector Level Weekend..A2-29

Figure 22 - Aircraft Present...A2-30

Figure 23 - Arrivals / Departures...A2-31

Figure 24 - Sectorisation ANOVAR – Working Position............................A2-32

Figure 25 - 'Communications' Factors / Working Position - Weekdays.............,,...A2-33

Figure 26 - 'Communications' Factors / Working Position – Weekends..................A2-33

Figure 27 - Percentage Communication Times - Weekdays..............................A2-34
Figure 28 - Percentage Communication Times - Weekends..............................A2-35
Figure 29 - 'Subjective'/ Work Position – Weekdays..A2-36
Figure 30 - 'Subjective'/ Work Position – Weekends...A2-36
Figure 31 - ISA Value Sectors..,A2-37
Figure 32 - NASA TLX (Task Load Index) Sectors...A2-38
Figure 33 - TMA ANOVAR..A2-40
Figure 34 - TMA ANOVAR TMA Level Weekday..A2-41
Figure 35 - TMA ANOVAR TMA Level Weekend...A2-42
Figure 36 - TMA - Arrivals/Departures...A2-43
Figure 37 - TMA - Aircraft Present..A2-44
Figure 38 - TMA - Performance Variates Weekday...A2-45
Figure 39 - TMA - Performance Variates Weekend...A2-46
Figure 40 - Dmk Arrival and Departure Delays... A2-47
Figure 41 - Dmk Separation Errors and Missed Approaches......................... A2-48
Figure 42 - Hentzau / Zenda Delays..A2-49
Figure 43 - Hentzau / Zenda Separation / Missed Approaches......................A2-50
Figure 44 - TMA ANOVAR Working Position...A2-51
Figure 45 - 'Communications' Factors / Working Position - Weekdays.........A2-52
Figure 46 - 'Communications' Factors / Working Position - Weekends........A2-53
Figure 47 - Strelsau TMA Communications /DUMMKOPF..............................A2-54
Figure 48 - Hentzau / Zenda Communications..A2-55
Figure 49 - TMA Subjective / Work Position Weekdays...................................A2-57
Figure 50 - TMA Subjective / Work Position Weekends...................................A2-58
Figure 51 - Strelsau Work Position/DUMMKOPF ISA.......................................A2-59
Figure 52 - Hentzau / Zenda Work Position ISA..A2-60
Figure 53 - Strelsau NASA TLX (Task Load Index)..A2-61
Figure 54 - Hentzau / Zenda NASA TLX...A2-62

Abbreviations

ATC	Air Traffic Control
ANOVAR	Analysis of Variance
CARD	Conflict and Risk Display
DUMMKOPF	Digital Universal Monitoring and Mediation Kontextual Operational Planning Function.
EEC	EREHWONTROL Experimental Centre, Utopia City, Utopia
EREHWONTROL	The Erehwon Organisation for the Safety of Air Navigation
FL	Flight Level
ICAO	International Civil Aviation Organisation
ISA	Instantaneous Self-Assessment
NASA	National Aeronautics and Space Agency
NASA-TLX	NASA Task Load Index
RCCATI	Ruritanian Civil and Commercial Aviation Training Institute

Executive Summary

In October 2020, a real-time simulation of the Ruritania airspace was carried out at the Erehwontrol Experimental Centre at Utopia City Utopia. The intention was to estimate the capacity increase to be expected by dividing the Ruritania airspace into two sectors, and to evaluate a semi-automatic approach sequencing function DUMMKOPF (Digital Universal Monitoring and Mediation Kontextual Operational Planning Function) - developed by Dr V. Frankenstein of the University of Strelsau. On the basis of objective and subjective measurements, it appears that the current airspace organisation should be adequate for approximately 150% of current 95% peak traffic on Weekdays, and 140% of current 955 peak traffic at Weekebds. A two-sector layout should cope with about 20% more traffic.

Application of DUMMKOPF. at Strelsau. International showed that it was not a significant improvement in its current form. Although the principle appears sound, some practical ATC considerations have not been taken into account, and the code supplied contained some undetected errors, affecting the system as a whole.

Unexpectedly, it appears that weekend traffic at Zenda. and particularly Hentzau. Airports will be the limiting factor to the growth of air traffic in Ruritania..Traffic is particularly heavy for Morning Weekend Arrivals, and Evening Weekend Departures.

Some relief may be achieved by re-scheduling Ruritanian Civil and Commercial Aviation Training Institute activity or moving it from Hentzau to other airports.

Temporary flow control restrictions on peak days may be used to reduce peak traffic, but would not reduce controller workload.

Controller opinion supports the observation that the current Ruritanian ATC system is considerably out of date, requiring updating of both equipment and methodology.

The simulation was completed on time and within budget. Action is already under way to implement its recommendations.

Chapter 1 - Introduction

Background

Ruritania. is a nation state situated in the western part of the continent of Erehwon, bordered on the north by Kennaquhair, on the east by Datong, on the south by Nephelocccygia and on the west by Atlantis. Ruritania is the main producer and exporter of coprolite, increasingly in demand for international relations.

The main airport is Strelsau. International. Ruritania. is a member of Erehwontrol, which coordinates the Air Traffic Control services of the nations of Western Erehwon. In view of increasing international traffic, and tourism, the airspace of Ruritania is becoming overloaded, and some action must be undertaken to relieve the controllers.

The Ruritanian Ministry of Aviation is considering splitting Ruritanian airspace into two sectors instead of the one existing. They are concerned that the additional co-ordination workload may balance out the reduction in traffic under control.

In addition, the Ministry is interested in evaluating a computer-based sequencing aid for departures and arrivals at Strelsau. International, avoiding conflicts with local traffic from Hentzau and Zenda airports. The procedure, developed by the well-known Artificial Intelligence expert, Dr V. Frankenstein, needs to be validated.

At a meeting of the Erehwontrol steering committee, it was agreed that a real-time simulation should be mounted at the Erehwontrol Experimental Centre (EEC) to evaluate the proposed airspace re-organisation and the introduction of the Digital Universal Monitoring and Mediation Kontextual Operational Planning Function (DUMMKOPF).

Organisation of this report

Chapter 1 is this Introduction.

Chapter 2 lists the questions raised.

Chapter 3 provides a brief description of Ruritanian airspace.

Chapter 4 describes the experimental plan, with the proposed measurements.

Chapter 5 describes the running of the simulation.

Chapter 6 summarises and discusses the analysis.

Chapter 7 repeats the questions raised, with the conclusions drawn. It also includes some additional conclusions.

The report finishes with acknowledgements.

Chapter 2 – Questions Raised

- Will the proposed division of Ruritanian airspace provide increased capacity during normal week-day operation?

- Will the proposed division of Ruritanian airspace provide increased capacity during week-end operation?

- Is the Digital Universal Monitoring and Mediation Kontextual Operational Planning Function (DUMMKOPF.) an acceptable function for Ruritanian Airspace?

Chapter 3 – Ruritanian Airspace

General Map

Figure 1 presents a simplified general view of Ruritania airspace as currently divided.

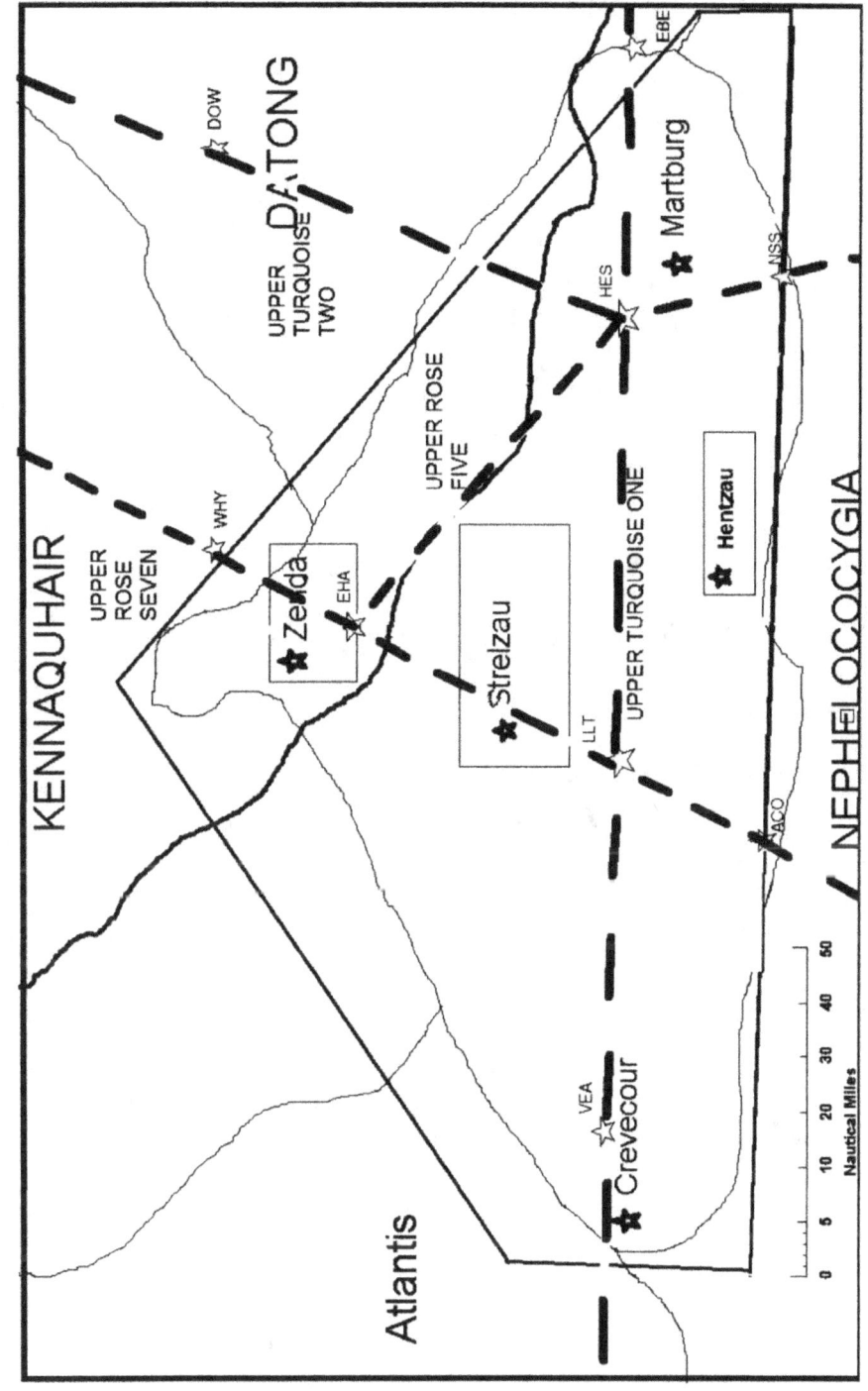

Figure 1 - Ruritania Airspace

Weekday Traffic

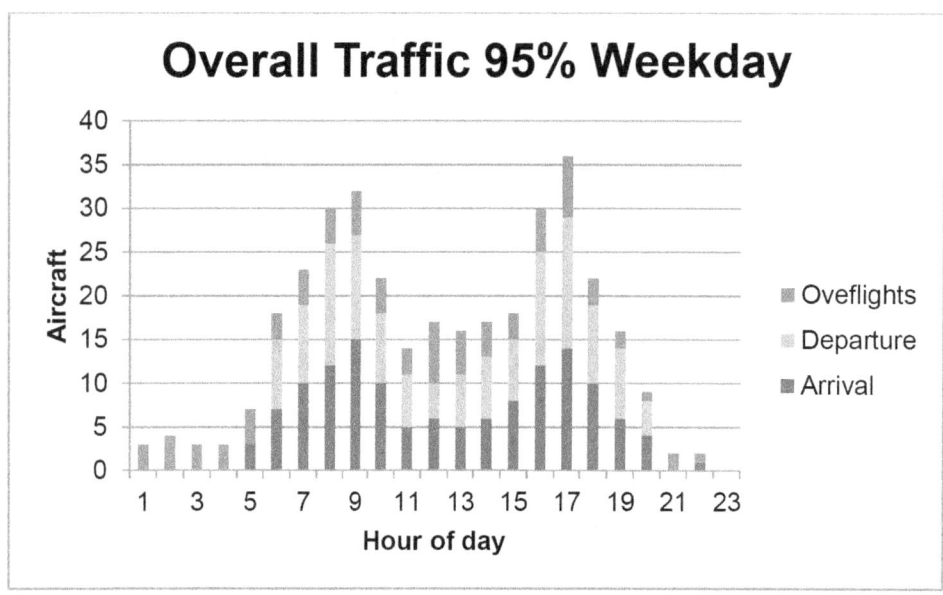

Figure 2 Overall Traffic 95% Weekdays

Figure 2 presents the 95th percentile traffic arriving, departing and overflying Ruritanian Airspace on weekdays.

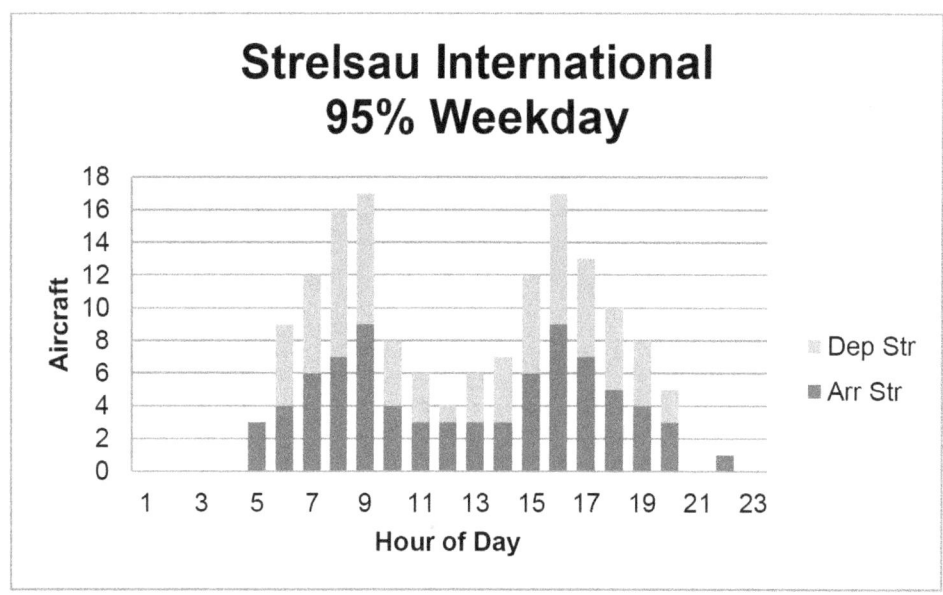

Figure 3 - Strelsau International 95% Weekday Traffic

Figure 3 presents the hourly arrival and departure rates for Strelsau. International for the 95%ile traffic level weekday traffic.

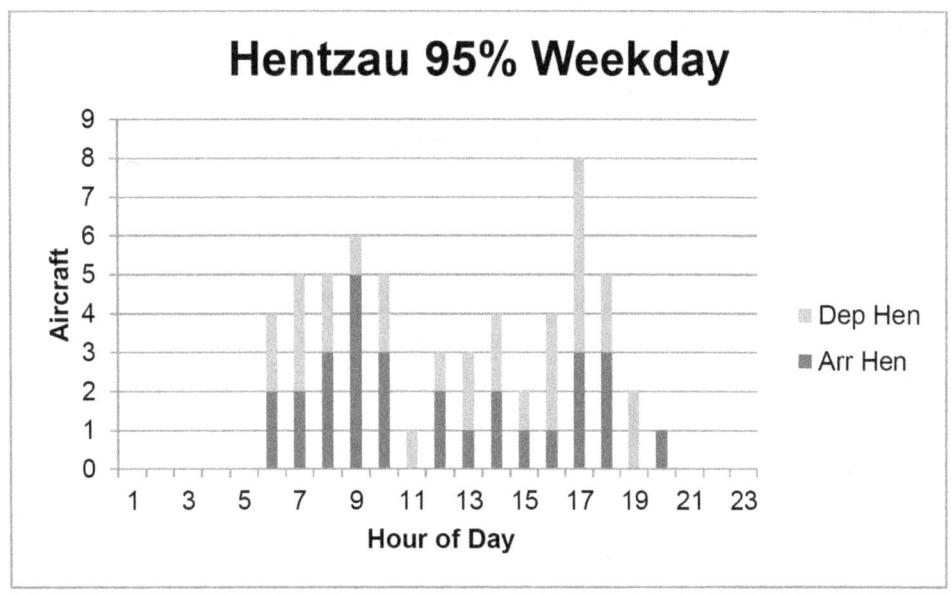

Figure 4 - Hentzau 95% Weekday Traffic

Figure 4 presents the hourly arrival and departure rates for Hentzau. Airport for the 95%ile traffic level weekday traffic.

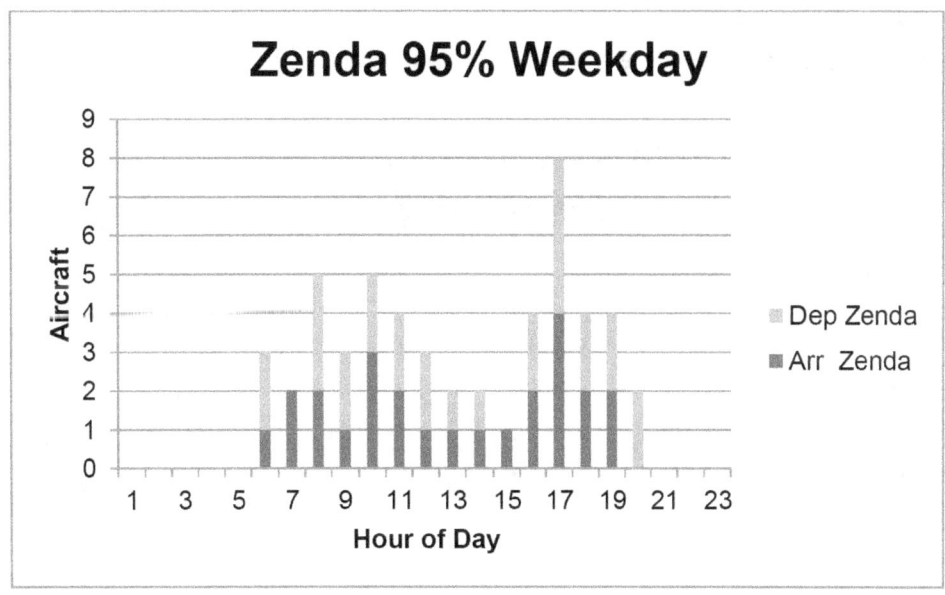

Figure 5 - Zenda 95% Weekday Traffic

Figure 5 presents the hourly arrival and departure rates for Zenda. Airport for the 95%ile traffic level weekday traffic. Figures 2-5 show that the peak hours for weekly traffic occur from 0800-1000 and 1600-1800 for all three major airports in Ruritania.

Weekend Traffic

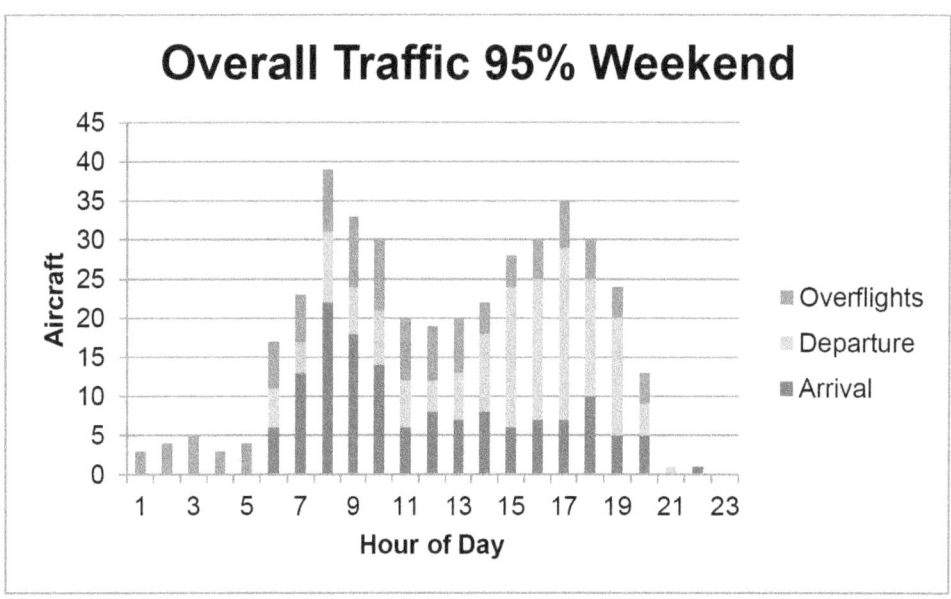

Figure 6 - Overall Traffic 95% Weekend

Figure 6 presents the average hourly arrival, departure and overflight rates for Ruritanian Airspace, during the 95%ile traffic level weekend traffic.

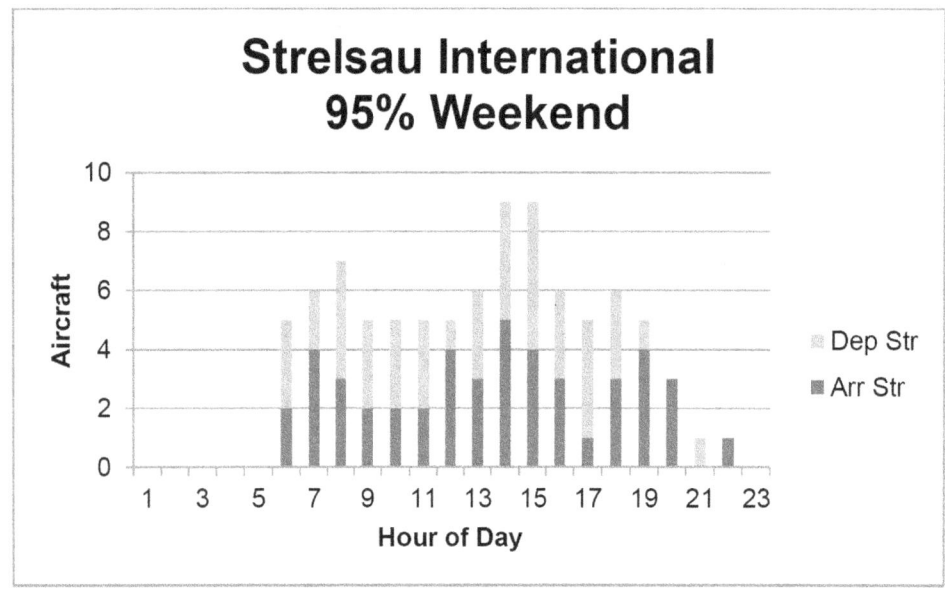

Figure 7 - Strelsau International 95% Weekend

Figure 7 presents the hourly arrival and departure rates for Strelsau International airport during the 95%ile traffic level weekend traffic.

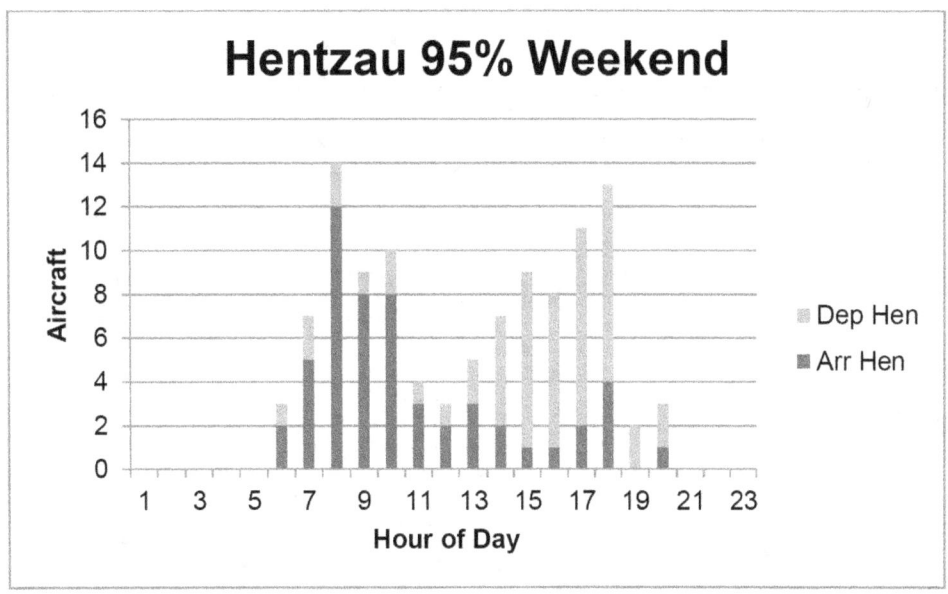

Figure 8 - Hentzau 95% Weekend traffic

Figure 8 presents the hourly arrival and departure rates for Hentzau airport during the 95%ile traffic level weekend traffic.

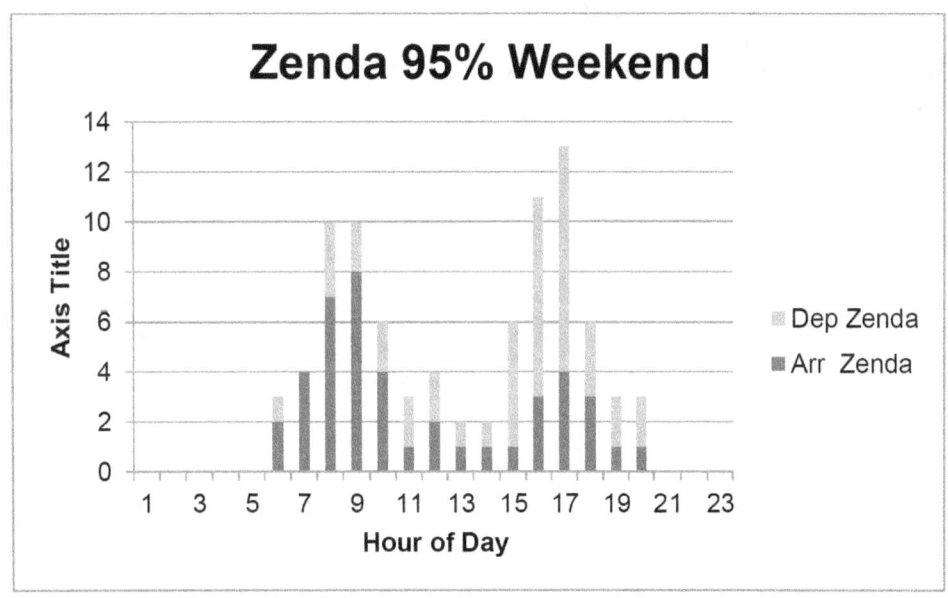

Figure 9 - Zenda 95% Weekend Traffic

Figure 9 presents the hourly arrival and departure rates for Zenda airport during the 95%ile traffic level weekend traffic. Figures 6 to 9 show that the weekend traffic pattern is very different from the weekday pattern. Strelsau International is relatively quiet, but both Hentzau and Zenda experience peaks of arrivals from 0800 – 1000 and of departures from 1600 – 1800.

Airspace Organisations

Figure 10 presents the initially proposed split into Ruritania East and Ruritania West.

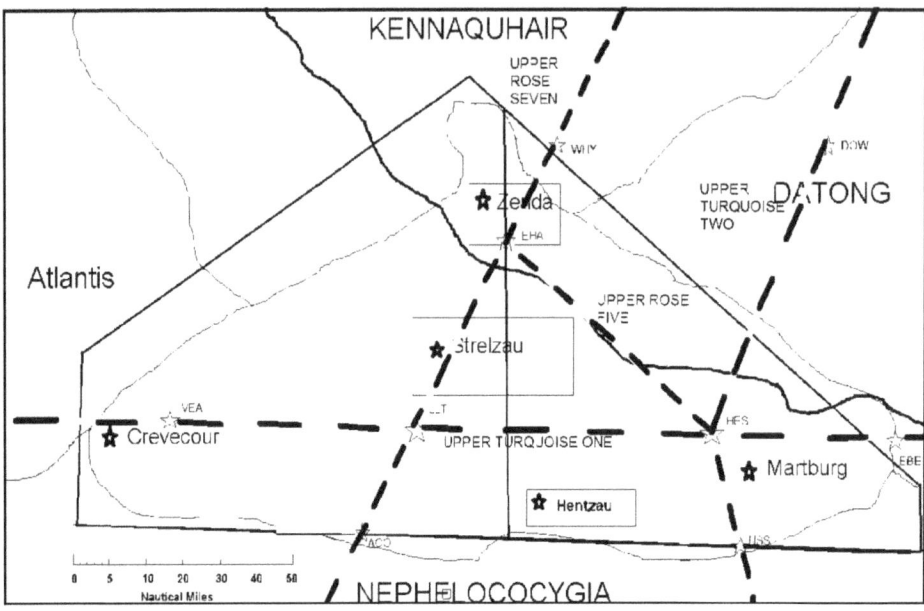

Figure 10 - Ruritania Airspace - Two Sector Initial Proposal

Figure 11 presents a slightly modified split into Ruritania East and Ruritania West, taking into account some ATC considerations.

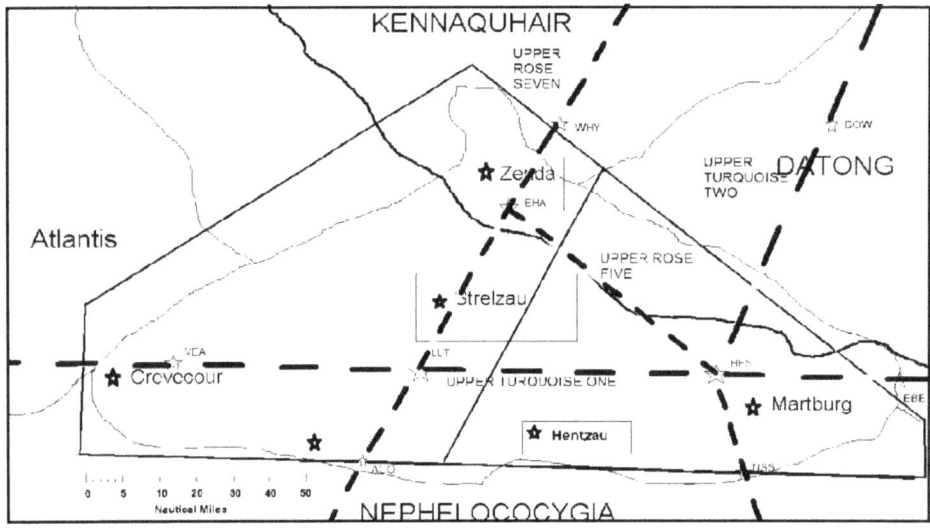

Figure 11 - Ruritania Airspace - Two Sector Revised Proposal

Traffic samples

Samples of existing traffic for the peak daily periods of 0800-1000 hrs and 1600-1800 hrs for 95 percentile traffic on weekdays and weekends were extracted by Ruritania Control for the year 2019. These were checked for minor abnormalities and compared with the base of aircraft types available at Erehwontrol Experimental Centre.

The verified samples were then doubled by the repetition of each flight at a an interval of 5 to 15 minutes, with a modified identity, preserving the airline identity, but varying the flight number.

Samples were derived from this data base by selecting 50, 80, 100, 125, 150 and 175 per cent samples. A preliminary check was made to detect entry conflicts and adjust entry times accordingly. Potential conflicts within the controlled area were retained. This process was repeated each time a sample was required, so that no traffic sample was presented twice, although many flights were in common, reflecting the real situation.

Participating Staff

Two teams of ten qualified, current controllers were selected to take part in the simulation. All controllers completed the simulation, except for one who was taken ill during the last session of the last week.

Chapter 4 – Experimental Plan and Measurements

Experimental Plan

Figure 12 explains the characteristics of each simulation run, as summarised in its name.

Thus the first exercise, A1WkAMX50% taking place at 1400 hours on Monday 12 October, involved

- Team A,
- using a single-sector layout,
- with weekday morning traffic
- using DUMMKOPF
- at 50% of current density.

This was the first training exercise, and was intended to familiarise the participants with the control room and the procedures involved, including ISA, NASA/TLX and DUMMKOPF.

Characteristic	Value	Value
Team	A = First Team	B = Second Team
No of Sectors	1 = One Sector	2 = Two Sector
Time of Day	AM = Morning	PM = Afternoon
Time in Week	Wk = Weekday	We = Weekend
DUMMKOPF	X = Used	(blank) Not Used
% Current traffic	50/80 Training	100/125/150/175 Measured

Figure 12 - Coding of Simulation Conditions

Figure 13 shows the experimental plan as agreed before the commencement of the simulation.

Control Room Simulation

Day of Week/ Session	Monday 12/10 Training	Tuesday 13/10 Training	Wednesday 14/10 Measured	Thursday 15/10 Measured
0900-1100	Arrival	A2WkAM- 80%	B1WkAM- X 100%	A2WkPM- X 100%
1100-1300	Introduction	B2WkPM- 80%	A1WkAM- X 100%	B2WkPM- X 100%
1400-1600	A1WkAM- X 50%	A2WeAM- 80%	B1WePM- X 100%	A2WeAM- X 100%
1600-1800	B1WkPM- X 50%	B2WePM- 80%	A1WePM- X 100%	B2WeAM- X 100%

Day of Week/ Session	Monday 19/10 Measured	Tuesday 20/10 Measured	Wednesday 21/10 Measured	Thursday 22/10 Measured
0900-1100	B1WkAM- 125%	A2WkPM- X 125%	B1WkAM- 150%	A2WkPM- X 150%
1100-1300	A1WkAM- 125%	B2WkPM- X 125%	A1WkAM- 150%	B2WkPM- X 150%
1400-1600	B1WePM- 125%	A2WeAM- X 125%	B1WePM- 150%	A2WeAM- X 150%
1600-1800	A1WePM- 125%	B2WeAM- X 125%	A1WePM- 150%	B2WeAM- X 150%

Day of Week/ Session	Monday 26/10 Measured	Tuesday 27/10 Measured	Wednesday 28/10 Measured	Thursday 29/10 Measured
0900-1100	B1WkAM- 175%	A2WkPM- X 175%	B1WkAM- 100%	A2WkPM- X 100%
1100-1300	A1WkAM- 175%	B2WkPM- X 175%	A1WkAM- 100%	B2WkPM- X 100%
1400-1600	B1WePM- 175%	A2WeAM- X 175%	B1WePM- 100%	A2WeAM- X 100%
1600-1800	A1WePM- 175%	B2WePM- X 175%	A1WePM- 100%	B2WeAM- X 100%

Figure 13 - Simulation Experimental Plan

Expert Opinion

ISA (Instantaneous Self-Assessment)

During all exercises the well-known ISA (Instantaneous Self-Assessment) measure was taken at 3- minute intervals. Training exercises and the first fifteen minutes of each exercise and any measurements after 90 minutes were not included in the analysis.

ISA is applied through a computer-based system. Each operator is supplied with a simple keyboard with five keys. At regular three-minute intervals the keys begin to flash. They continue until the operator pushes one of the keys, or for 20 seconds. The five keys represent the operator's self-assessment of his workload at that time. In the EREHWONTROL system the five levels are marked with the symbols and colours shown in Figure 14 below: -

Level	Symbol	Colour	Definition
VERY HIGH	+ +	Red	Completely occupied, some tasks missed
HIGH	+	Amber	Almost completely occupied, but task can be completed
FAIR	=	Green	Steady, reasonable workload. Some breaks
LOW	-	Cyan	Little work, much spare time
VERY LOW	- -	Magenta	Practically no work, boredom, lack of stimulus

Figure 14 - ISA Values

The ISA responses and the time taken to make them are recorded for each position, and the current vales are displayed at the supervisor's position.

NASA-TLX (NASA Task Load Index)

The equally well known NASA-TLX (NASA Task Load Index) was applied after each exercise. The data for training exercises was not analysed formally, although the responses were taken into account in deciding when training could be considered adequate and measured exercises could start.

The NASA-TLX is applied immediately after each simulation run. It requires the operator to rank his performance on six scales as defined in Figure 15 below: In the full application, each operator is required to rank the scales in order of importance and a weighted mean is derived. In practice, each scale is of interest in itself, and the NASA-TLX combined is not used.

Mental Demand	How much mental and perceptual activity was required (e.g. thinking, deciding, calculating, remembering, looking, searching, etc.)? Was the task easy or demanding, simple or complex, exacting or forgiving?
	LOW |_|_|_|_|_|_|_|_|_|_|_|_|_|_|_|_|_|_|_| HIGH
Physical Demand	How much physical activity was required (e.g. pushing, pulling, turning, controlling, activating, etc.)? Was the task easy or demanding, slow or brisk, slack or strenuous, restful or laborious?
	LOW |_|_|_|_|_|_|_|_|_|_|_|_|_|_|_|_|_|_|_| HIGH
Temporal Demand	How much time pressure did you feel due to the rate or pace at which the tasks or task elements occurred? Was the pace slow and leisurely or rapid and frantic?
	LOW |_|_|_|_|_|_|_|_|_|_|_|_|_|_|_|_|_|_|_| HIGH
Perform-ance **Careful!**	How successful do you think you were in accomplishing the goals of the task set by the experimenter (or yourself)? How satisfied were you with your performance in accomplishing these goals?
	GOOD |_|_|_|_|_|_|_|_|_|_|_|_|_|_|_|_|_|_|_| **POOR**
Effort	How hard did you have to work (mentally and physically) to accomplish your level of performance?
	LOW |_|_|_|_|_|_|_|_|_|_|_|_|_|_|_|_|_|_|_| HIGH
Frustration Level	How insecure, discouraged, irritated, stressed and annoyed versus secure, content, relaxed and complacent did you feel during the task?
	LOW |_|_|_|_|_|_|_|_|_|_|_|_|_|_|_|_|_|_|_| HIGH

Figure 15 - NASA-TLX Scales

Additional Subjective measures

In addition to these two formal instruments, a free form questionnaire was provided after each exercise, and de-briefings were held immediately after each exercise. As far as possible, the team carrying out a second exercise was separated from the first team, although it is impossible, and may be undesirable, to prevent controllers from discussing their experiences. A final end-of-simulation questionnaire was also issued and collected on the final day of the simulation, when a general overall de-briefing was carried out.

(For long-term development, a user-characteristics questionnaire was provided, to be filled on a voluntary basis. This forms part of a continuing study of controllers demography, background and physical wellbeing, which will be reported separately when completed.)

System Measures

These measures describe the run as a whole, taken from the records stored during the run.. A short name is attached to each measure, for future reference. Although data are usually recorded as individual values, they are usually expressed over short time intervals. A three-minute interval was chosen. Because traffic builds up at the start of a simulation run, the first fifteen minutes of traffic was discarded.

The following variates were derived for three-minute intervals.

The number of aircraft entering, leaving or present in a sector is called Ac, and specified as entering, leaving and present.. The exact definition of ' Aircraft Present' is the mean number of aircraft present during the defined interval, not the number of different aircraft present at any time during the interval.

Number of aircraft entering	EAc
Number of aircraft leaving	LAc
Number of aircraft present	PAc

The 'frequency' is the simulated R/T frequency, usually one per sector.

Number of Frequency Calls	Nfr
Duration of Frequency Calls as % of time	Tfr

The 'Intercom' is the simulated link between working positions within the simulated area – this reflects the movement of aircraft from one sector to another or a TMA..

Number of Intercom Calls	Nin
Total duration of Intercom Calls as % of time	Tin

'Handovers' are calls to and from the Feed sectors.

Number of Handovers	Nho
Total time of Handovers as % of time	Tho

'Orders' are the instructions entered into the computer system

Number of Orders	Nor
Total Time spent entering orders as %age of time	Tor

Control Room Simulation

No of Incomplete or Cancelled orders	NXo
Duration of Incomplete or Cancelled orders	TXo

For each TMA, some measures of their efficiency were recorded.

Number of aircraft landing	NLa
Number of aircraft departing	NDe
Seconds delay landing	DLa
Seconds delay departing	DDe

As well as being presented to the exercise supervisor, the ISA responses and the time taken to respond were also recorded at three minute intervals.

ISA response at three minute intervals	VISA
Time to respond to ISA query	TISA

(For each variate measured at three-minute intervals the mean values is also recorded, using the values from 15 to 90 minutes only, since the first fifteen minutes are unrepresentative because the system starts with no aircraft present. These values have the suffix M.. Thus MEAc represents the mean no of aircraft entering.)

Some values are recorded for the total measured part of the simulation run. This is usually because they correspond to events so rare that there are too few to provide reliable three-minute values

The number of 'mild' conflicts – Aircraft approaching within 5 NM – and of 'severe' conflicts – Aircraft approaching within 2 NM were recorded for the period of the simulation. For TMAs the number of Missed Approaches was also recorded.

Mild Conflicts	NC5
Severe Conflicts	NC2
Missed Approaches	NMA.

In addition, the NASA-TLX scores were included' (The actual NASA Task Load Index was not calculated.)

Mental Demand	TLMe
Physical Demand	TLPh
Temporal Demand	TLTe
Performance	TLPe
Effort	TLEf
Frustration Level	TLFr

While all these data are recorded, those that are measured at three-minute intervals were recorded in a file for each exercise, clearly marked with exercise code, date and time. An additional file cumulated the exercise-length measures (TLX and Mean values) for each exercise, with the exercise code, date and time.

Choice of Measures

A very experienced Head of the Simulation Division at EEC, the legendary Mr. Charles Massie, devised a simple tabulation where the available methods were given on the Y-axis, and the aims of the simulation on the X-axis.

Examples of the Massie Grid are presented as Figures 16 and 17 following. In Figure 16 all available variates are listed. A question mark shows that this method may be relevant to this aim. An X shows that it is not. In Figure 17 and subsequently, only relevant variates are listed, to save space and for simplicity..

Sectorisation

We have two separate questions, the capacity increase and the use of DUMMKOPF,

We will therefore carry out one set of analyses for overall, planners and executives (Figure 16) relating to Sectorisation. and another set of analyses comparing the exercises using DUMMKOPF and similar exercises where DUMMKOPF is not used. (Figure 17).

The second and third columns refer to the weekday and weekend traffic, where the differences between the measures are important, but their actual values are often just as important. After all, if the number of intercom calls is twice as high in the two-sector case, it may be statistically significant, but sufficiently low that it does not matter.

The fourth column in Figure 16 shows which measures are for the whole sector, and which apply to each working position.

Measure	1 vs 2 Sectors Weekday	1 vs 2 Sectors Weekend	Sector / Work Posn
Mean of 20 Three-minute intervals			
EAc	?	?	Sector
Lac	?	?	Sector
PAc	?	?	Sector
NFr	?	?	Work Posn
TFr	?	?	Work Posn
Nin	?	?	Work Posn
Tin	?	?	Work Posn
NHo	?	?	Work Posn
THo	?	?	Work Posn
Nor	?	?	Work Posn
Tor	?	?	Work Posn
NXo	?	?	Work Posn
Txo	?	?	Work Posn
VISA	?	?	Work Posn
TISA	?	?	Work Posn
Overall Measures			
NLa	X	X	Sector
DLa	X	X	Sector
NDe	X	X	Sector
DDe	X	X	Sector
NC5	?	?	Sector
NC2	?	?	Sector
TLX	X	X	Work Posn
TLMe	?	?	Work Posn
TLPh	?	?	Work Posn
TLTe	?	?	Work Posn
TLPe	?	?	Work Posn
TLEf	?	?	Work Posn
TLFr	?	?	Work Posn

Figure 16 - Massie Grid - Sectorisation

Fifteen variates at 3 minute intervals and 8 overall measures were used for the comparison of one sector and two sector layouts. Each 3-minute variate supplies one overall mean value. There were, in all, 23 variates. Of these 5 refer to the sector, and 18 to individual working positions.

A2-23

TMA

For the three TMAs, with two working positions each, 18 means of three-minute measures and 10 overall measures. produce 28 variates for analysis. Although we are only really interested in Strelsau, where DUMMKOPF was installed, the analyses were repeated for Hentzau and Zenda – where the presence or absence of DUMMKOPF at Strelsau is unlikely to make any significant difference.. (The average of all three TMAs is of no interest to anyone.)

Measure	DUMMKOPF On/Off	TMA / Work position
Mean of 20 Three-minute Intervals		
EAc	?	TMA
Lac	?	TMA
PAc	?	TMA
NFr	?	Work position
TFr	?	Work position
Nin	?	Work position
Tin	?	Work position
NHo	?	Work position
THo	?	Work position
Nor	?	Work position
Tor	?	Work position
NXo	?	Work position
Txo	?	Work position
NLa	?	TMA
DLa	?	TMA
NDe	?	TMA
DDe	?	TMA
VISA	?	Work position
TISA	?	Work position
Overall Measures		
NC5	?	TMA
NC2	?	TMA
NMA	?	TMA
TLMe	?	Work position
TLPh	?	Work position
TLTe	?	Work position
TLPe	?	Work position
TLEf	?	Work position
TLFr	?	Work position

Figure 17 - Massie Grid - TMAs

Chapter 5 – Conduct of the Simulation

The Ruritania I simulation took place from Monday 12 October 2020 to Friday 30 October 2020, as planned. Four simulation runs were planned for each day on Monday to Thursday of each week, Fridays being reserved for re-runs where a planned exercise could not be carried out, or where technical problems were encountered.

Monday 12th and Tuesday 13th were devoted to settling in and training exercises. Measured exercises were carried out on Wednesday 14th and Thursday 15th.Two measured exercises were not completed during week 1, so were repeated on Friday 16th morning. Considerable technical problems were encountered with DUMMKOPF.

Four exercises were run on each day from Monday 19th to Thursday 22nd. One exercise in week 2 was not completed due to problems with DUMMKOPF, and was re-run on Friday 23/10, with an exercise in Week 1 where data was not properly recorded.

A party from the Ruritania Civil and Armed Forces Aviation (Marshal Strakenz and Colonel Sapt from the Ruritanian Army, Captain Von Tarlenheim from the Ruritanian Air Force and Monsieur Ronde de Cuir from the Ruritanian Air Ministry) visited the Centre on Thursday 22nd. They were briefed on the history and capabilities of the EEC, and its contribution to air safety and economy. They viewed a simulation in progress, and spoke to the participating controllers, to their mutual satisfaction..

Four exercises were planned for each day from Monday 26th to Thursday 29th October. However, two exercises were lost on Monday 26/10 am, owing to an external power failure. The remaining exercises were rescheduled to minimise disruption, requiring two exercises on the morning of Friday 30/10. All; four exercises involving one sector at 175% of current peak traffic were terminated prematurely, as the system was clearly overloaded. One exercise involving two sectors at 175% was also terminated for the same reason. The final exercises, involving a repeat of the original 100% traffic level (with different traffic – see Chapter 3) were completed satisfactorily. Two exercises were run on Friday 30th Morning, completing the simulation within the scheduled time.

All participating controllers cooperated enthusiastically with the simulation staff, and carried out their duties in an exemplary manner. (One controller experienced a non-work-related injury on Thursday 29th afternoon, but the team was able to compensate for the disturbance.)

Chapter 6 – Analysis

Strategy

The data collected have been listed in Chapter 4 above. In addition, the conclusions of each session debrief, participants' immediate comments, and the overall questionnaires, individual comments and final de-briefing have been considered.

Two initial technical points should be made.

1. Subjective responses to the ISA measure are on an Ordinal scale, which should be treated differently from continuous measurements.

2. Successive measurements at three-minute intervals are rarely statistically independent. For example the ISA levels reported by some controllers are identical from 15 to 90 minutes. Assuming independence would lead to exaggerated levels of significance.

The pragmatic solution is to assume that ordinal measures are equally scaled, with unit difference between levels, and to analyse only the mean values for each exercise. This produces a considerable reduction in the sheer mass of data to be handled. In this simulation, we were not concerned with the changes in stress or strain within each exercise, as one might be if considering reactions to sudden changes in workload, such as might be associated with emergencies. (Emergencies were not simulated on this occasion.)

In addition, it will be noted that five of the eight 175% traffic exercises were cancelled as the controllers were clearly overloaded. Accordingly, the 175% traffic level was omitted from the analyses.

100% traffic samples were run twice. During the set of exercises at the start learning effects appeared to be present, and considerable difficulty was experienced with DUMMKOPF. Accordingly, the final 100% level exercises were included in the analyses

In all, 24 measured exercises were included in the analysis. The two major questions for this simulation concern the effects of revised Sectorisation and the introduction of DUMMKOPF to the Strelsau TMA. These are virtually independent and can be analysed separately.

Sectorisation – Factors – Sector level

Figure 13 (above) shows the planned experimental design. Figure 12 lists the potential factors. Although it is usual to extract the effects of as many factors as possible, there are some that will be irrelevant to the analysis of Sectorisation effects.

Considering the potential factors in sequence:

The **Team** is not relevant to the analysis of sector effects

The **Sectorisation** is very important – this is the primary question for the simulation. However, we have records for one sector (RU) for the one sector case, and for two sectors (RUE and RUW) for the two sector case. We can separate the two degrees of freedom between sectors into

1. NoS (**Number of Sectors**) contrasting the single sector RU with the average of the two sectors (RUE and RUW).
2. REW (**RURITANIA East vs West**) contrasting the RUE sector with the RUW sector..

The **Time of Day** and **Day of Week** are important – if there are significant differences overall, or relative to the **Number of Sectors**. Because the Weekday/Weekend capacities are given as the main questions, this factor is really a separator – we analyse Weekdays separately from Weekends.

The presence or absence of **DUMMKOPF** is not relevant to the Sectorisation,

The **Percentage of Current traffic** is relevant, since capacity differences are more likely to matter when the workload is heavy.

Figure 16, the Massie Grid for Sectorisation, includes a column listing the variates that are in common for a given sector (such as the number of aircraft present) or that differ for different working positions (such as the ISA value.)

Figure 18 shows the relevant factors, with a short abbreviation and the alternative levels involved.

Factor	Code	Value 1	Value 2	Value 3
No of Sectors	**NoS**	1 = One Sector	2 = Two Sector	
East vs West	**REW**	1 = RUE	2 = RUW	
Time of Day	**ToD**	AM = Morning	PM = Afternoon	
% Current traffic	**Tra**	100	125	150

Figure 18 - Sectorisation ANOVAR Factors

Figure 19 lists the factors and the potentially relevant interactions for sector variates. (For example the NoS **x Tra** interaction may show that the two sector layout has an advantage over the one sector layout under heavy traffic, but not under light traffic.)

A full report would require a tabulation for 5 variates x 2 ANOVARs (weekday/weekend) x 9 factors - 90 tabulations in all. These can be supplied on request but in the interests of simplicity, only 'significant' interactions are reported, and illustrated graphically or by tabulation.

Factor	Degrees of Freedom
NoS	1
REW	1
ToD	1
Tra	2
NoS x ToD	1
NoS x Tra	2
REW x ToD	1
REW x Tra	2
TOD x Tra	2
Residual	4
TOTAL	17

The sensitivity of the ANOVAR depends on the number of residual degrees of freedom. Non-significant interactions are therefore folded into the residual, the least significant first, until either there are no more non-significant interactions, or only the main factors are left, whether they are significant or not significant at the 5% level

In the following tables, three stars represent a difference significant at the 0.1% level. Two stars represent a difference significant at the 1% level. One star represents a difference at the 5% level. . A blank cell indicates no significant difference.

To economise on space, interactions that are not significant for any variate – a completely blank line - are omitted. Equally, variates that are not affected significantly by any factor are omitted.

Figure 19 - Sectorisation Anovar - Sector Level

Control Room Simulation

Figure 20 summarises the analyses (Figure 19) for weekday traffic of the 5 Sector level variates, (Figure 16) giving the significance of remaining differences.

Factor	EAc	Lac	Pac
NoS	*	*	**
REW	*	*	*
Tra	**	**	**
ToD		*	*
NoS x Tra	*	*	*

Figure 20 - Sector Level Weekday

Figure 21 summarises the analyses (Figure 19) for weekend traffic of the 5 Sector level variates, (Figure 16) giving the significance of remaining differences.

Factor	EAc	Lac	Pac
NoS	*	*	**
REW	*	*	*
Tra	**	**	**
ToD	**	**	
NoS x Tra	*	*	*

Figure 21 - Sector Level Weekend

Figures 22 and 23 present the significant differences in the mean numbers of aircraft present, and of aircraft arriving and departing during the measured hour.

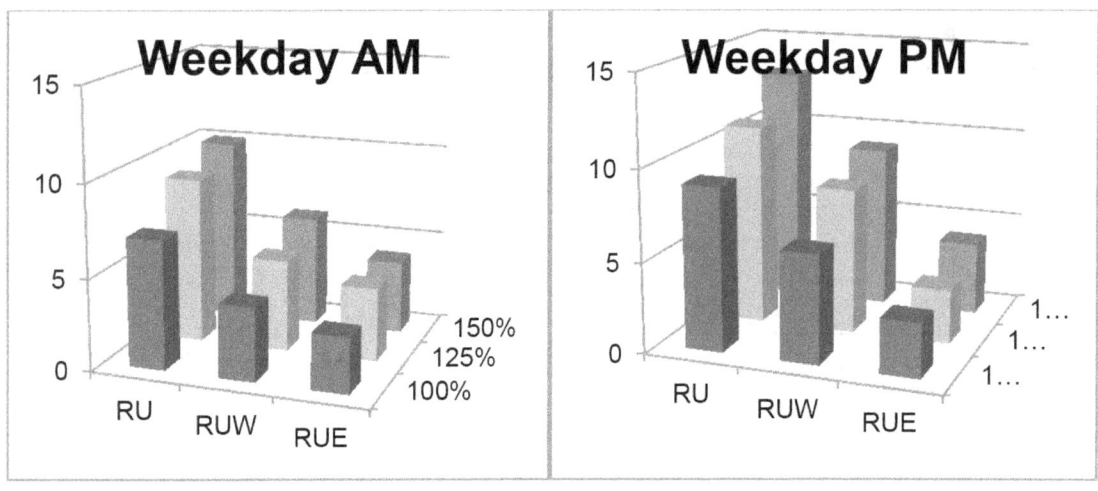

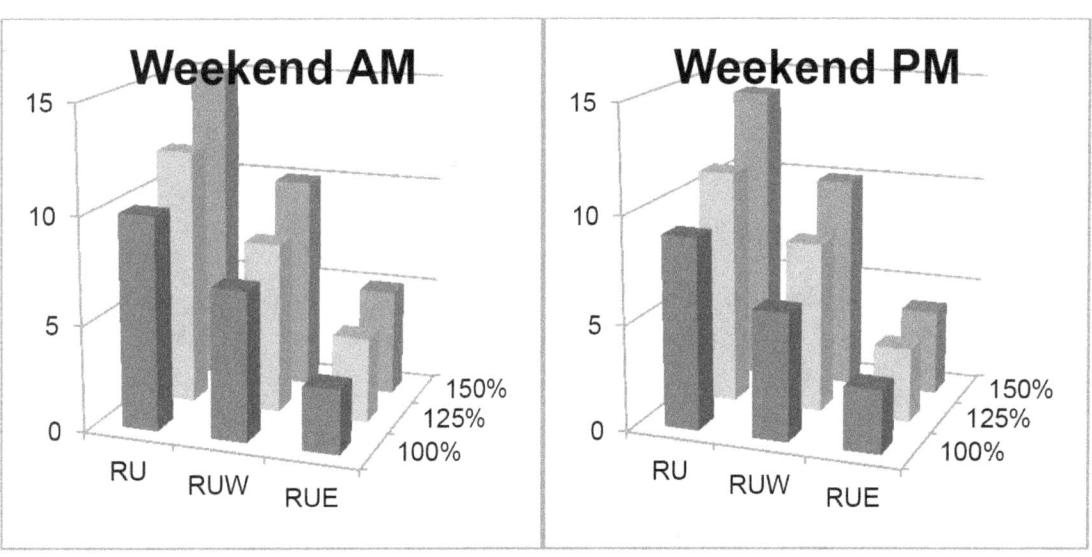

Figure 22 - Aircraft Present

Figure 22 shows the mean number of aircraft present by sector for increasing traffic load. The total of traffic for RUE and RUW is equal, within experimental limits, to the traffic for RU. This is as expected. RUW has significantly more traffic than RUE, because at least two of the three principal airports must be in one sector, rendering an even division of traffic impossible. The increases in aircraft present are proportional to the percentages selected.

Figure 23 show that arrivals and departures follow a generally similar pattern, with an even distribution between arrivals and departures on weekdays, with about 75% of RU for RUW and about 40% of RU for TUW – the additional numbers representing the effect of subdivision. At Weekends, the traffic is definitely tidal, with a predominance of arrivals ia the morning and departures at the evening peaks.

In short, the traffic behaved as required, and was representative of current peak traffic.

A2-30

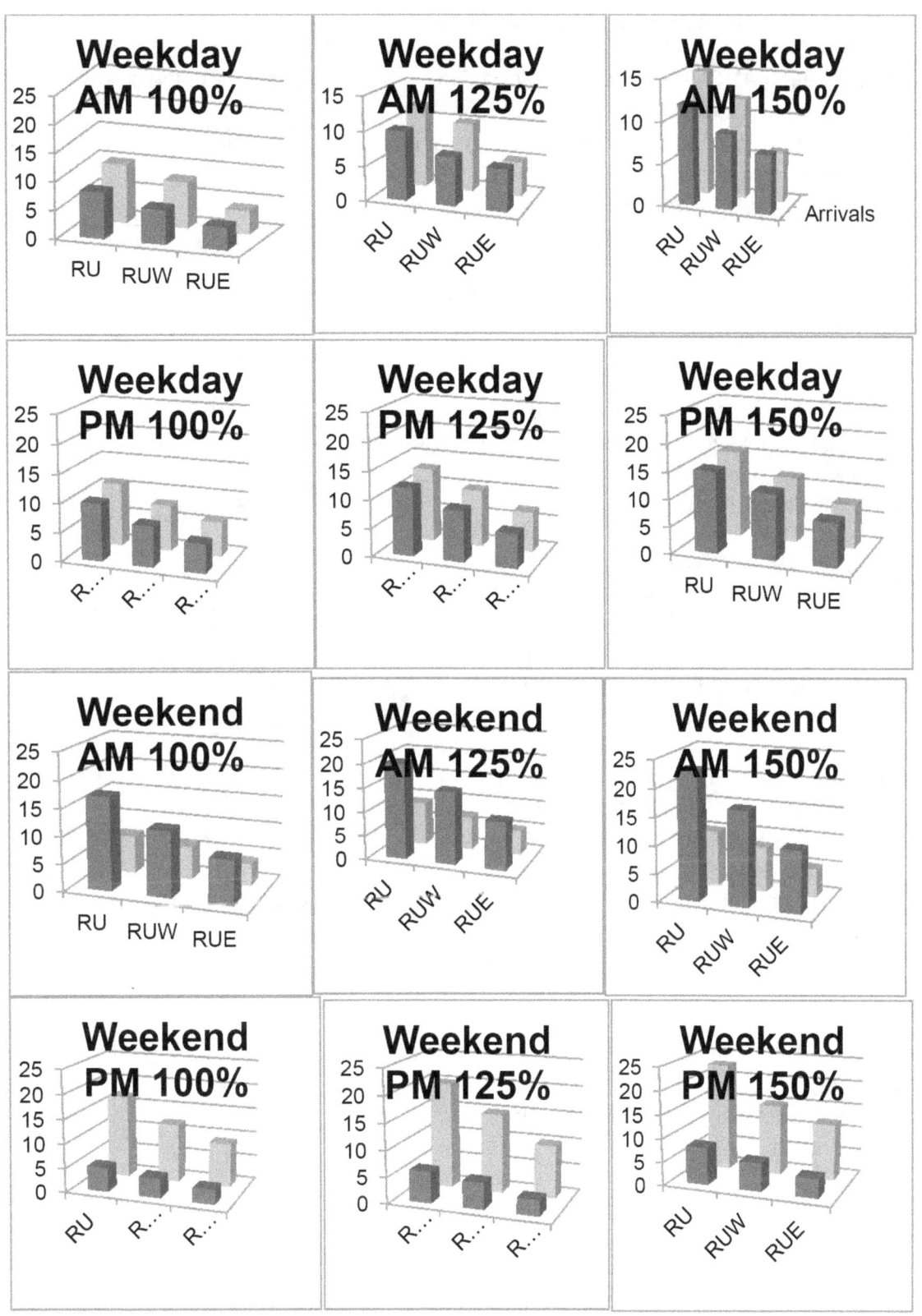

Figure 23 Arrivals / Departures (Arrivals Front – Departures Back)

Sectorisation – Factors Working Position Level

The four Sectorisation ANOVAR factors shown in Figure 18 are the factors between exercises, as shown in the exercise title. In addition, there will be separate files for the planner(s) and executive(s) within the exercises, giving an additional factor - **WORKING POSITION (WkP)** with a value of 1 for the Planner and 2 for the executive,

Figure 24 shows the 17 potentially significant interactions at the working position level.

A full report would require tabulation for 18 variates x 2 anovars (weekday/weekend) x17 factors - 612 tabulations in all. These can be supplied on request but in the interests of simplicity, only 'significant' interactions are reported, and illustrated graphically or by tabulation.

The sensitivity of the ANOVAR depends on the number of residual degrees of freedom. Non-significant interactions are therefore folded into the residual, the least significant first, until either there are no more non-significant interactions, or only the main factors are left, whether they are significant or not.

In the following tables, three stars represent a difference significant at the 0.1% level. Two stars represent a difference significant at the 1% level. One star represents a difference significant at the 5% level. A blank cell indicates no significant difference. To economise on space, interactions that are not significant for any variate – a completely blank line - are omitted. Equally, variates that are not affected significantly by any factor are omitted.

Because there are so many variates involved, the factor/ variate tabulation has been split into two parts – the first containing 'communications" measures, and the second 'subjective' (ISA and NASA TLX)

Factor	Degrees of Freedom
NoS	1
REW	1
ToD	1
Tra	2
WkP	1
NoS x ToD	1
NoS x Tra	2
NoS x WkP	1
REW x ToD	1
REW x Tra	2
REW x WkP	1
ToD x Tra	2
ToD x WkP	1
Tra x WkP	2
NoS x ToD x Tra	2
NoS x ToD x WkP	1
ToD x Tra x Wkp	2
Residual	11
TOTAL	35

Figure 24 - Sectorisation ANOVAR - Working Position

Figure 25 summarises the analyses (figure 24) for weekday traffic of the 8 working position level 'communications' variates, (figure 16) giving the significance of remaining differences.

Factor	NFr	TFr	NIn	Tin	NHo	THo	NOr	TOr
NoS	**	**	**	**	**	**	**	**
Tra	**	**	**	**	**	**	**	**
WkP	***	***	**	**	**	**	**	**
NoS x Tra	*		*		*			*
NoS x WkP	**	**	**	**		**	**	**
Tra x WkP	*	*	*	*	**	*	*	**

Figure 25 - 'Communications' Factors/ Working Position - Weekdays

Figure 26 summarises the analyses (figure 24) for weekend traffic of the 8 working position 'communications' level variates, (figure 16) giving the significance of remaining differences.

Factor	NFr	TFr	NIn	Tin	NHo	THo	NOr	TOr
NoS	**	**	**	**	**	**	**	**
Tra	*		*	*	*	**	**	**
WkP	*		*		*	*	*	**
NoS x Tra	*		*	*	*	*	*	*
NoS x WkP	*	*	*				*	*
Tra x WkP		*		*		*	*	

Figure 25 - 'Communications' Factors/ Working Position - Weekends

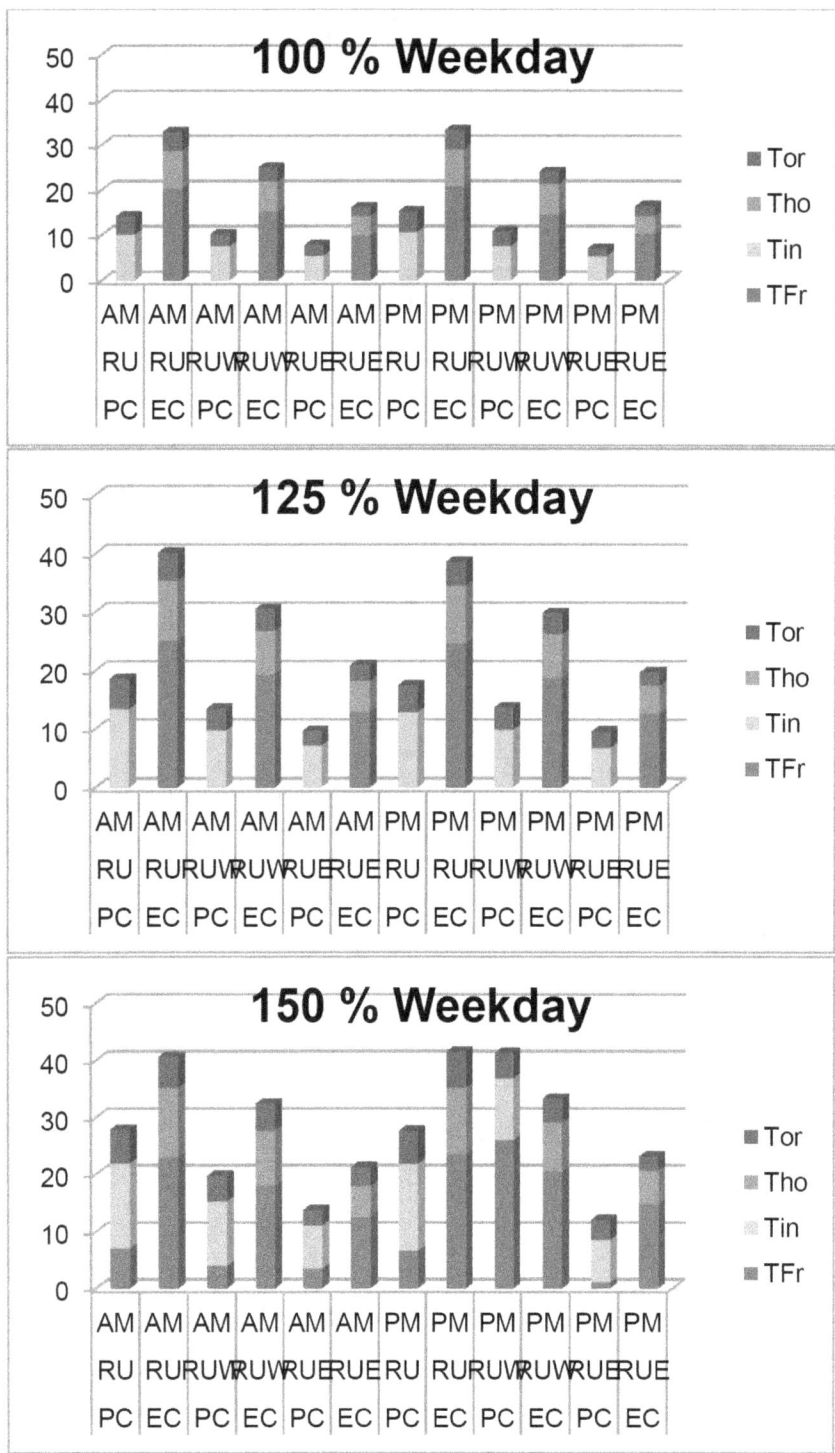

Figure 27 - Percentage Communications Times - Weekdays

Figure 27 illustrates the significant differences in the percentages of communication times on the frequency, of Intercom communications, of handover procedures and control orders, for sector working positions, by sector and time of day of week for each level of traffic for weekdays.

Control Room Simulation

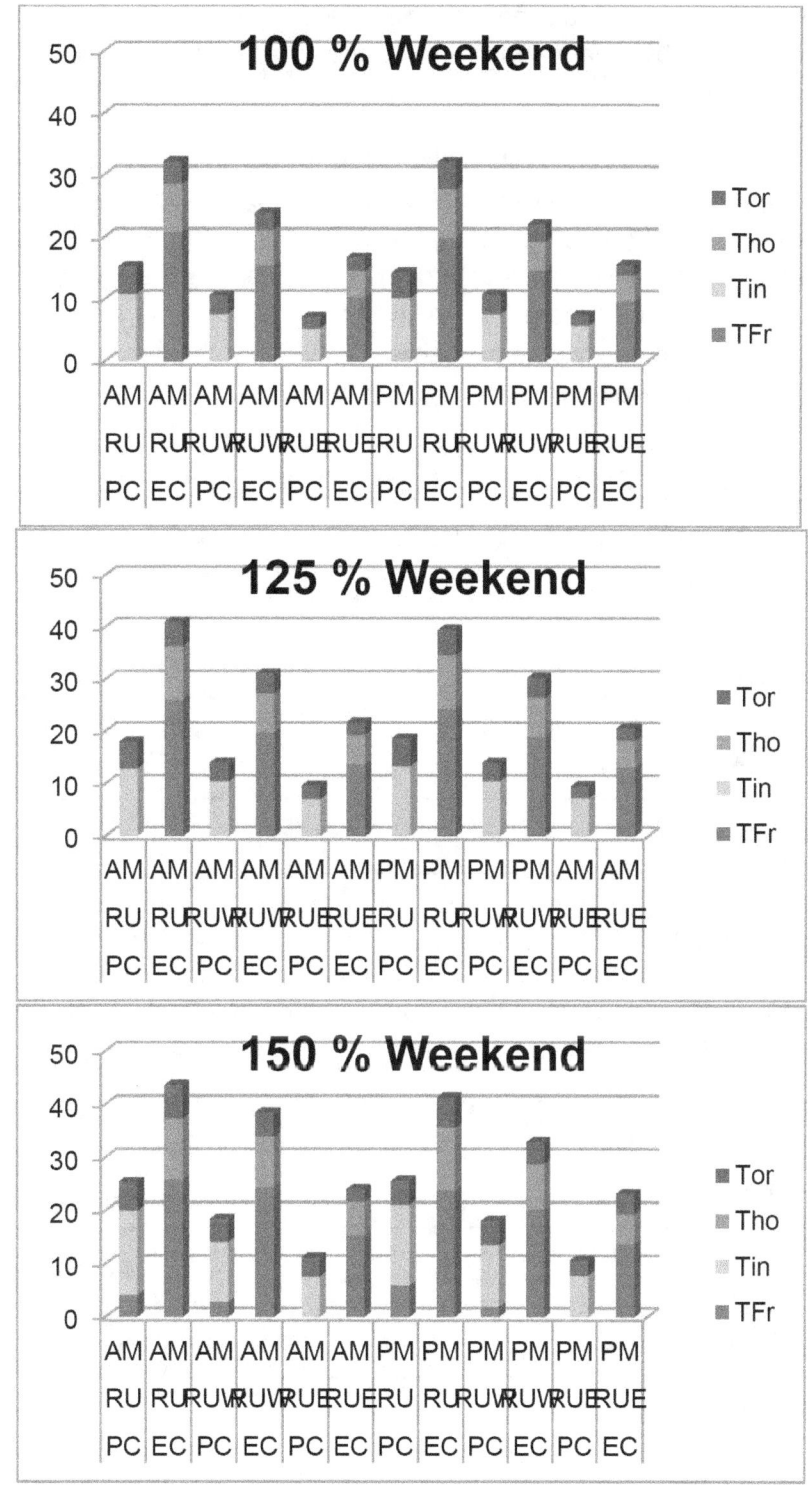

Figure 28 - Percentage Communications Times - Weekends

Figure 28 illustrates the significant differences in the percentages of communication times on

A2-35

the frequency, of Intercom communications, of handover procedures and control orders, for sector working positions, by sector and time of day of week for each level of traffic for weekends

Figures 27 and 28 show that the communication load for the executive controller in the single-sector layout is approaching what is usually considered saturation for 150% of the current traffic. It also shows that the PCs are assisting the ECs at this time – something that should not happen, but invariably does. . The corresponding figures for numbers of communications are very similar and are not included here for brevity. Figure 29 summarises the analyses (figure 24) for weekday traffic of the 8 working position 'subjective' variates, (figure 16) giving the significance of remaining differences

Factor	VISA	TLMe	TLPh	TLTe	TLPe	TLEf	TLFr
NoS	**	**	**	**	**	**	**
Tra	*		*	*	*	**	**
WkP	*		*		*	*	*
NoS x Tra	*		*	*	*	*	*
NoS x WkP	*	*	*				*

Figure 29 - 'Subjective' / Work Position - Weekdays

Figure 30 summarises the analyses (figure 24) for weekend traffic of the 8 working position 'subjective' variates, (figure 16) giving the significance of remaining differences.

Factor	VISA	TLMe	TLPh	TLTe	TLPe	TLEf	TLFr
NoS	**	**	*	**	*	*	**
Tra	*		*	*	*	**	**
WkP	*		*		*	*	*
NoS x Tra	*		*	*	*	*	*
NoS x WkP	*		*				*
Tra x WkP		*		*		*	*

Figure 30 - 'Subjective' / Work Position - Weekends

Control Room Simulation

Figure 31 illustrates the mean value of ISA ((Instantaneous Self-Assessment) for each sector, Working Position and traffic level, separately for each Time of Day and Day of Week. The corresponding response times show no significant differences, and show a highly non-normal distribution – most values being of the order of 500 milliseconds, with a few of 15-20 seconds, and a few non-responses. The occurrence of long responses and of non-responses appears to be random, except for a brief period in one exercise from when the controller at one working position became disabled until a colleague took over the position.

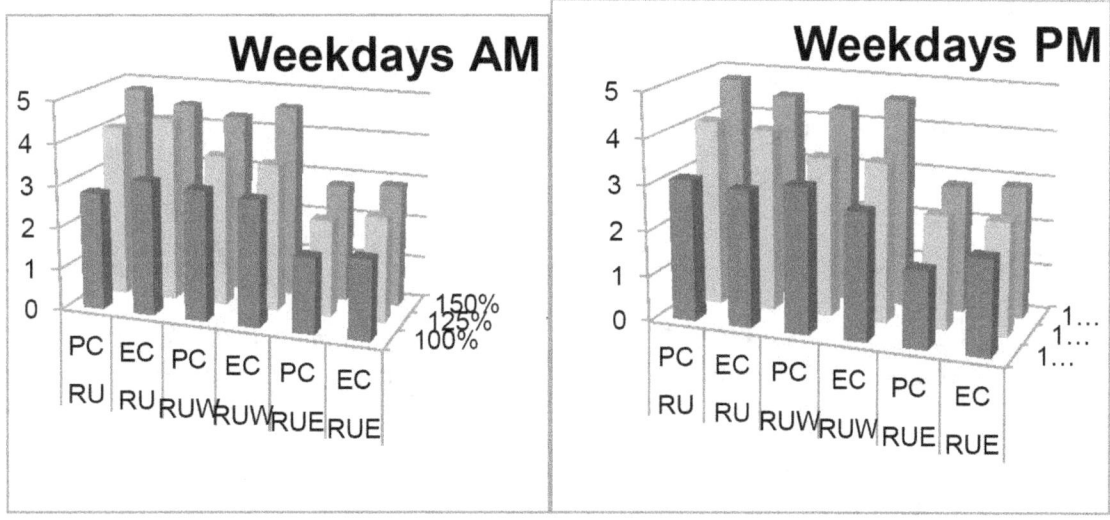

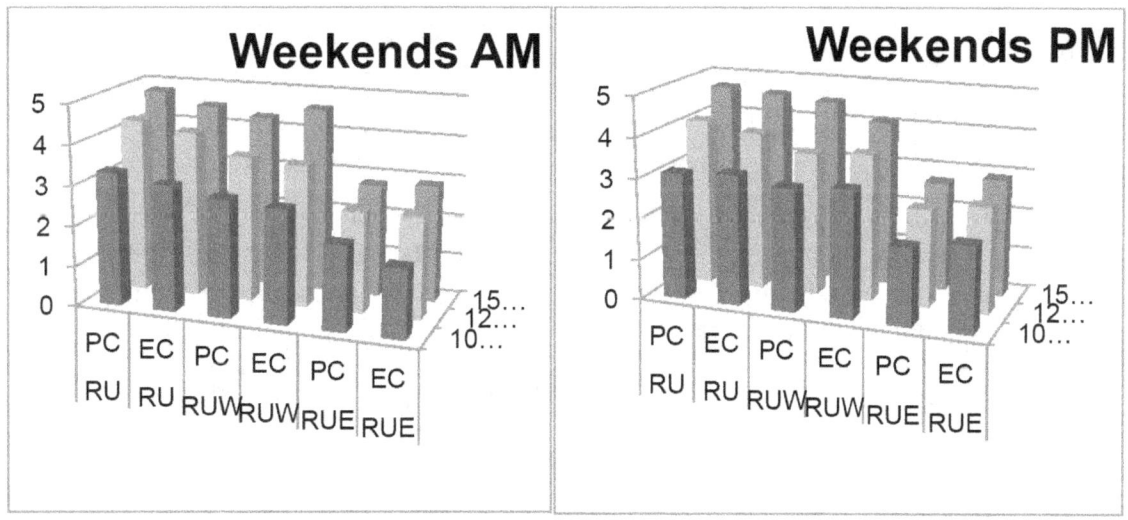

Figure 31 - ISA Values - Sectors

A2-37

Figure 32 illustrates the mean value of the six scales of the NASA TLX (Task Load Index) for each Sector, Working Position and Traffic Level - Time of Day and Day of Week do not differ significantly.

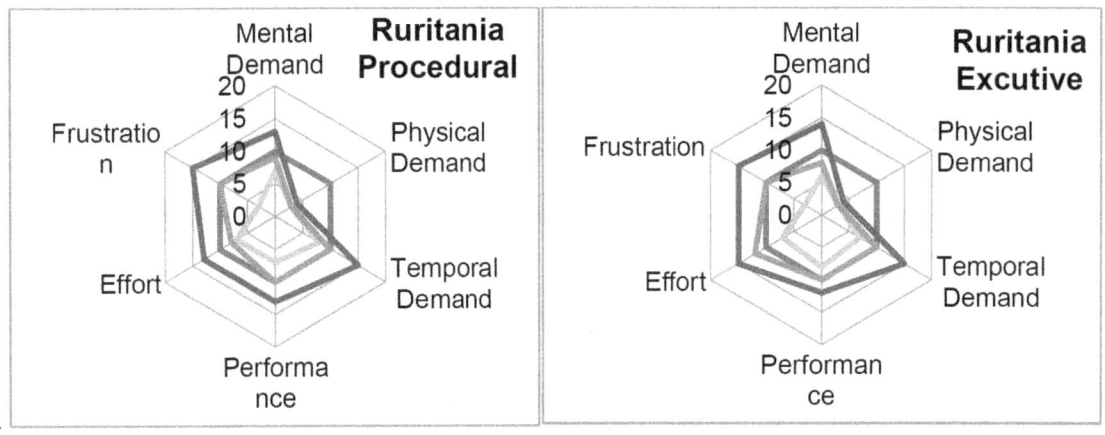

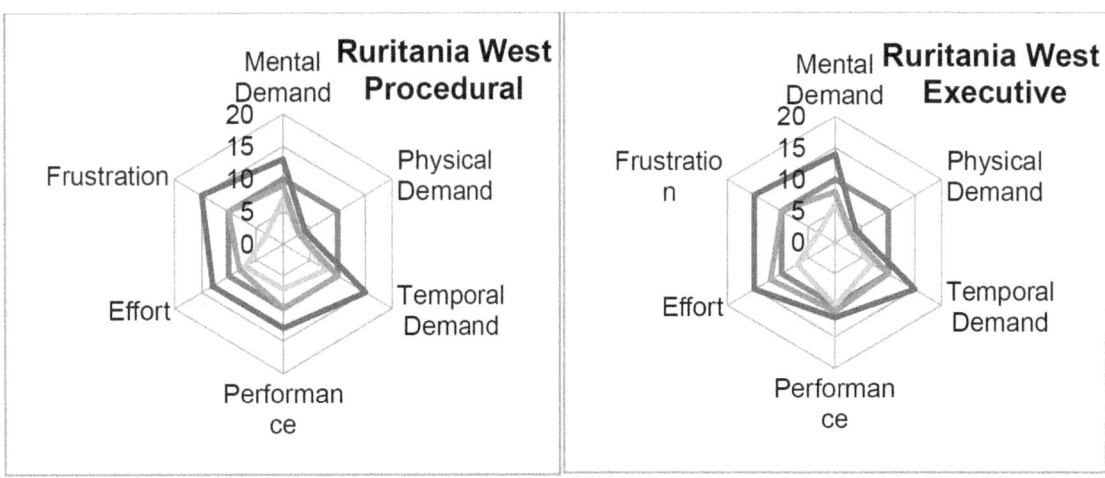

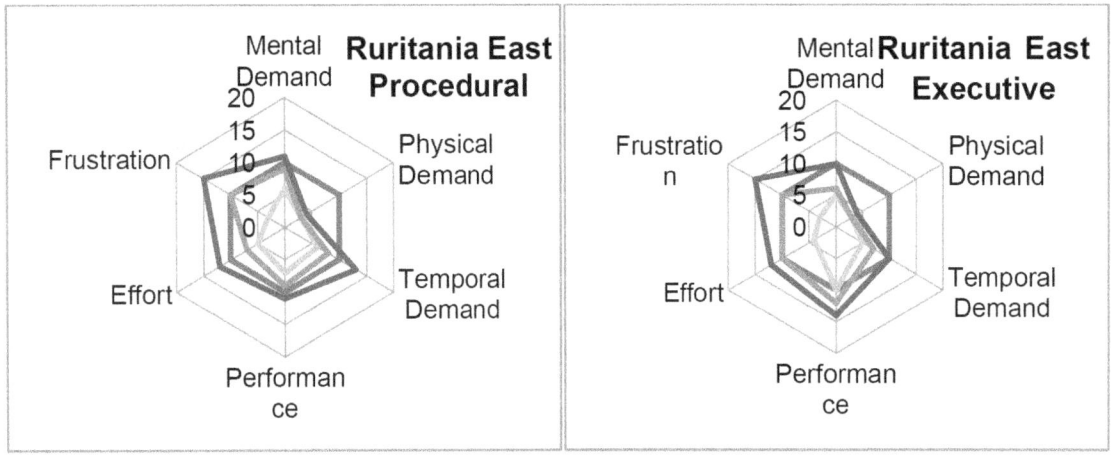

Figure 32 - Nasa TLX (Task Load Index) - Sectors

Sectorisation - Discussion

As was expected, the Single Sector layout involved significantly greater individual workloads than the Two Sector Layout. However, more detailed examination shows some interesting and unexpected features.

Given the existence of three major airports, it was not possible to divide the single Sector layout into two equal parts. Consequently, Ruritania West had a workload of about 70% of that of the Single sector, and Ruritania East had a relatively light workload of about 40% of Single Sector. (The extra 10% is generated by the interface between Ruritania East and Ruritania West.)

The Single Sector layout can cope well with the current (100%) 95%ile load, and with difficulty with 125%. It can handle 150% on weekdays,, but cannot safely handle 150%.at Weekends

The Two Sector layout would handle up to an estimated 170% traffic on Weekdays and 160% at Weekends.

Several additional points should be considered.

The traffic we are here considering is the 95% ile, occurring about 18 days in the year. Temporary traffic flow restrictions on these days may be acceptable for a few years.

Equally, these projections assume no significant change in the mixture of traffic in future years. Particularly at weekends, some traffic might be diverted from Hentzau, and to a lesser extent from Zenda. to Strelsau. international which is less loaded at weekends.

Additionally, it appears that a significant proportion of Hentzau. traffic originates with the Ruritanian Civil and Commercial Aviation Training Institute, flying short training flights. If these could be moved to Strelsau. International, or simply moved to off-peak times, a substantial reduction of peak traffic can be achieved.

Finally, It should be noted that the equipment and procedures at Ruritania Control are generally rather 'traditional' and that updating of equipment and procedures would be welcomed by the controllers, and provide increased capacity and safety.

TMA Analysis

The analysis for the performance of TMAs follows the same general strategy as for sectors, except that different factors are involved. In addition, the three TMAs are independent subjects of investigation, and are therefore subject to separate analysis.

Figure 17, the Massie Grid for TMAs, includes a column listing the variates that are in common for a given TMA (such as the number of aircraft present) or that differ for different working positions (such as the ISA value.) There are in all 10 variates that apply to the TMA as a whole, and 18 that refer to individual working positions.

TMA – Factors – TMA level

The ten TMA variates are : EAc,Lac,Pac, (aircraft entering, leaving or present), NLa and Dla (aircraft landing, mean delay in landing), NDe and DDe (aircraft departing and mean delay in departing), NC5,NC2 and NMA (Number of close approaches within 2NM, within 5 NM and number of Missed Approaches)

Figure 33 lists the factors and the potentially relevant interactions for TMA variates.

Factor	Degrees of Freedom
Dmk	1
ToD	1
Tra	2
Dmk x ToD	1
Dmk x Tra	2
TOD x Tra	2
Dmk x ToD x Tra	2
Residual	12
TOTAL	23

A full report would require a tabulation for 10 variates x 2 (weekday/weekend) anovars,x 3 TMAs x 7 factors - 420 tabulations in all. These can be supplied on request but in the interests of simplicity, only 'significant' interactions are reported, and illustrated graphically or by tabulation.

The sensitivity of the ANOVAR depends on the number of residual degrees of freedom. Non-significant interactions are therefore folded into the residual, the least significant first, until either there are no more non-significant interactions, or only the main factors are left, whether they are significant or not significant at the 5% level

In the following tables, three stars represent a difference significant at the 0.1% level. Two stars represent a difference significant at the 1% level. One star represents a difference at the 5% level. . A blank cell indicates no significant difference.

To economise on space, interactions that are not significant for any variate – a completely blank line - are omitted. Equally, variates that are not affected significantly by any factor are omitted.

Figure 33 - TMA ANOVAR

The analyses for each of the three TMAs are shown together to simplify the necessary explanation.

Control Room Simulation

Figure 34 summarises the analyses (Figure 33) for weekday traffic of the 3 TMA level Traffic variates, (Figure 17) giving the significance of remaining differences.

STRELSAU TMA			
Factor	EAc	LAc	PAc
ToD			*
Tra	**	**	**
ToD x Tra	*	*	

HENTZAU TMA			
Factor	EAc	LAc	PAc
ToD	*	*	*
Tra	**	*	**
ToD x Tra	*	*	

ZENDA TMA			
Factor	EAc	LAc	PAc
ToD	*	*	*
Tra	**	**	**
ToD x Tra	*	*	

Figure 34 - TMA ANOVAR TMA Level Weekdays

The factor **Dmk** (presence or absence of DUMMKOPF) does not affect the amount of traffic for any TMA.

Figure 35 summarises the analyses (Figure 33) for weekend traffic of the 3 TMA level Traffic variates, (Figure 17) giving the significance of remaining differences.

STRELSAU TMA			
Factor	EAc	LAc	PAc
ToD	*	*	*
Tra	**	**	**
ToD x Tra	*	*	

HENTZAU TMA			
Factor	EAc	LAc	PAc
ToD	*	*	*
Tra	**	**	**
ToD x Tra	*	*	

ZENDA TMA			
Factor	EAc	LAc	PAc
ToD	*	*	*
Tra	**	**	**
ToD x Tra	*	*	

Figure 35 - TMA ANOVAR TMA Level Weekends

The factor **Dmk** (presence or absence of DUMMKOPF) does not affect the amount of traffic for any TMA.

Figure 36 illustrates the effects of the Traffic Level and Time of Day for traffic arriving and departing for each TMA, for Weekdays and Weekends.

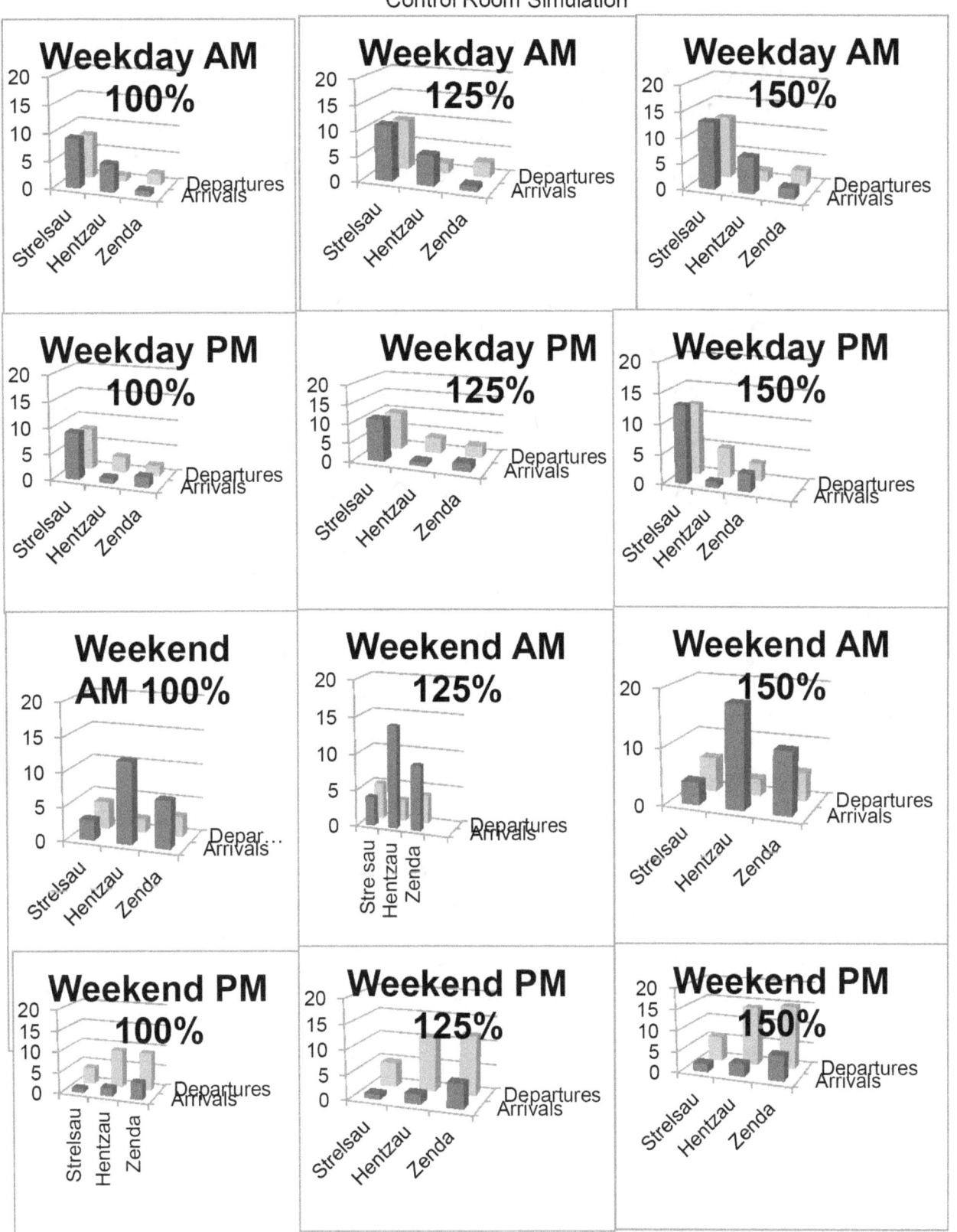

Figure 36 - TMAs Arrivals / Departures

Figure 37 illustrates the effects of the Traffic Level and the Time of Day for traffic present in each TMA, for weekdays and Weekends.

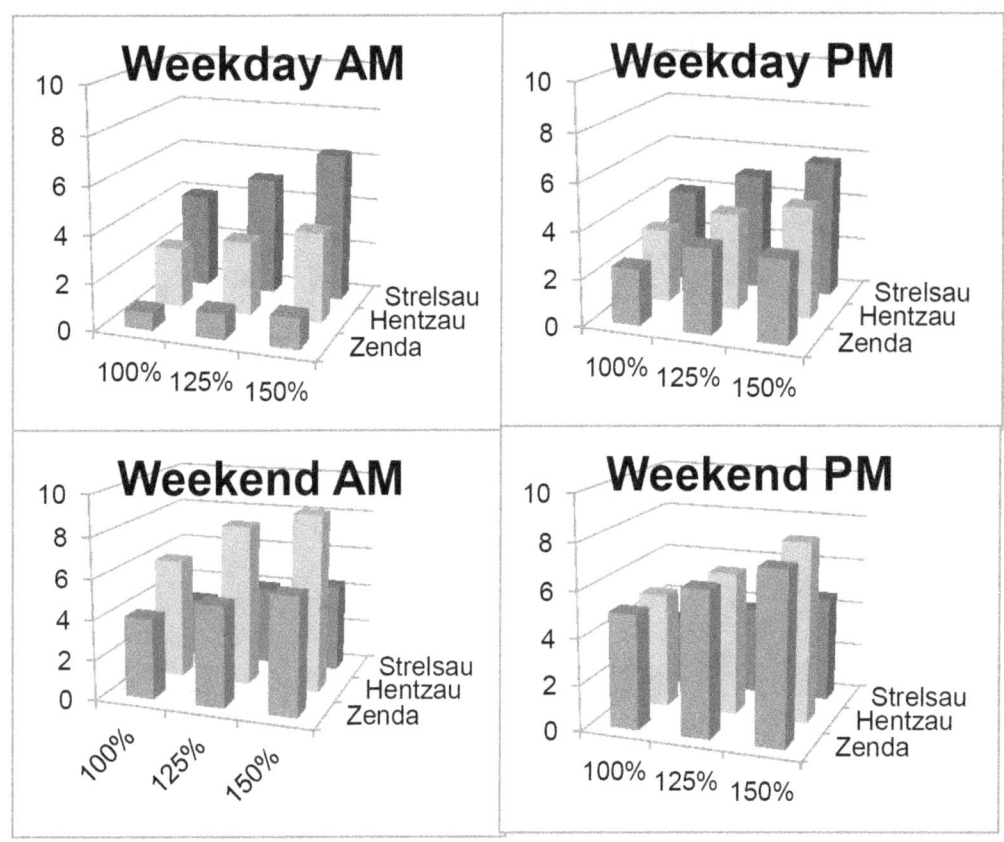

Figure 37 - Aircraft Present

Figures 36 and 37 confirm that the actual traffic conforms to that planned. They also show that Strelsau. TMA is more highly loaded than Hentzau or Zenda on weekdays, but that Strelsau is relatively lightly loaded at weekends, while both Hentzau and Zenda experience heavy arrival traffic in the morning peak and heavy departure traffic in the evening peak.

Figure 38 summarises the analyses (Figure 33) for weekday traffic of the 7 TMA level Performance variates, (Figure 17) giving the significance of remaining differences.

(for convenience these variates are :
DLa and DDe, the mean delay in landing and departure
NC5 the number of minor separation infringements, NMA the number of missed approaches
MLa and NDe are redundant, since they merely confirm that the traffic was as planned.
NC2 the number of major infringements is too small for meaningful analysis)

Strelsau TMA				
Factor	DLa	DDe	NC5	NMA
Dmk	*	*	*	*
Tra	*	*	*	
Dmk x Tra	*	*		*

Hentzau TMA				
Factor	DLa	DDe	NC5	NMA
ToD	*	*		*
Tra	**		*	*

ZendaTMA				
Factor	DLa	DDe	NC5	NMA
ToD	*	*		*
Tra	**		*	*

Figure 38 - TMA Performance Variates Weekdays

The factor **Dmk** (presence or absence of DUMMKOPF) does not affect the performance for Hentzau ar Zenda TMA. It does however significantly affect StrelsauTMA. There are too few major separation infringements to reach statistical significance with respect to any factor.

A2-45

Figure 39 summarises the analyses (Figure 33) for weekend traffic of the 7 TMA level Performance variates, (Figure 17) giving the significance of remaining differences.

STRELSAU TMA				
Factor	DLa	DDe	NC5	NMA
Dmk	*	*	*	*
Tra	*	*	*	
Dmk x Tra	*	*		*

HENTZAU TMA				
Factor	DLa	DDe	NC5	NMA
ToD	**	**		*
Tra	**	**	*	

ZENDA TMA				
Factor	DLa	DDe	NC5	NMA
ToD	**	**		
Tra	**	*	*	

Figure 39 - TMA Performance Variates Weekends

The factor **Dmk** (presence or absence of DUMMKOPF) does not affect the performance for Hentzau 0r Zenda TMA. It does however significantly affect Strelsau TMA. There are too few major separation infringements to reach statistical significance with respect to any factor.

Figure 40 summarises the effects of **Dmk** (the presence or absence of DUMMKOPF), with the effect of **TRA** (the level of traffic) for the Strelsau TMA, on the mean arrival and departure delays as recorded in tables 38 and 39, for weekday and weekend traffic..

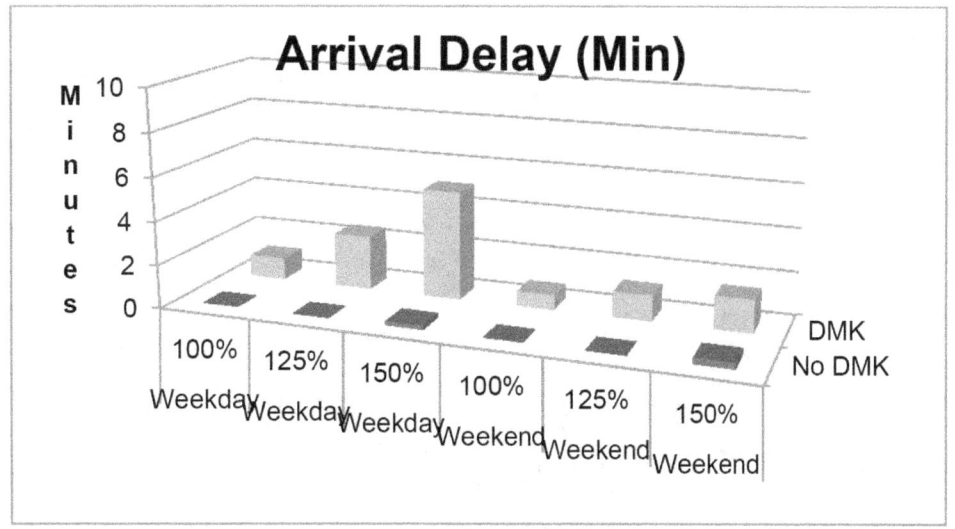

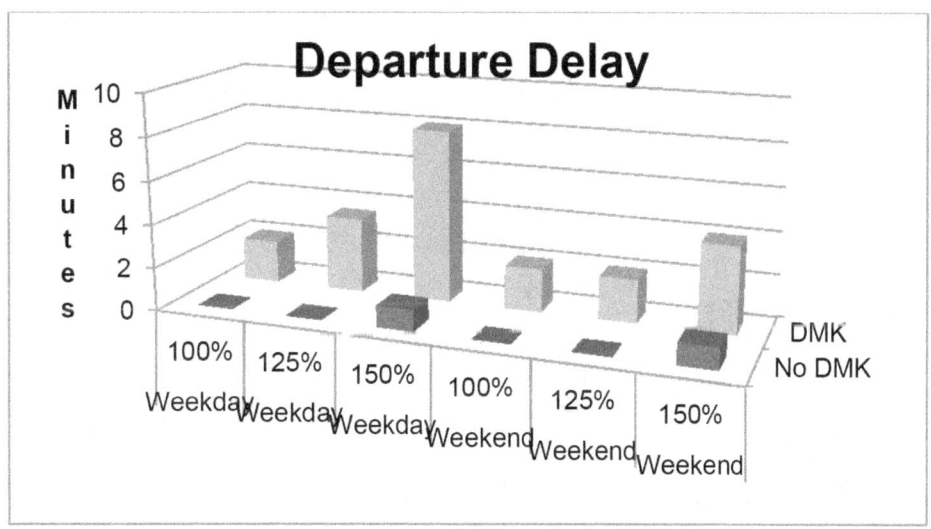

Figure 40 - Dummkopf Arrival / Departure Delays

figure 40 shows that arrival and departure delays are negligible without DUMMKOPF, but that they increase at least linearly with traffic load. The controllers allot delays evenly between arrivals and departures.

Figure 41 summarises the effects of **DMK** (the presence or absence of DUMMKOPF), with the effect oF **TRA** (the level of traffic) for the Strelsau TMA, on the mean number of minor separation errors and of missed approaches as recorded in tables 38 and 39, for weekday and weekend traffic..

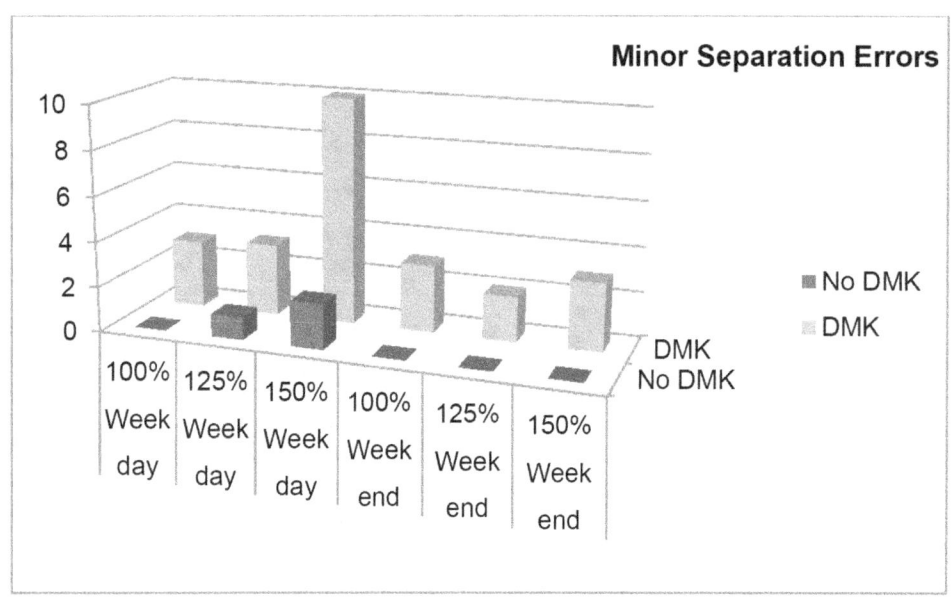

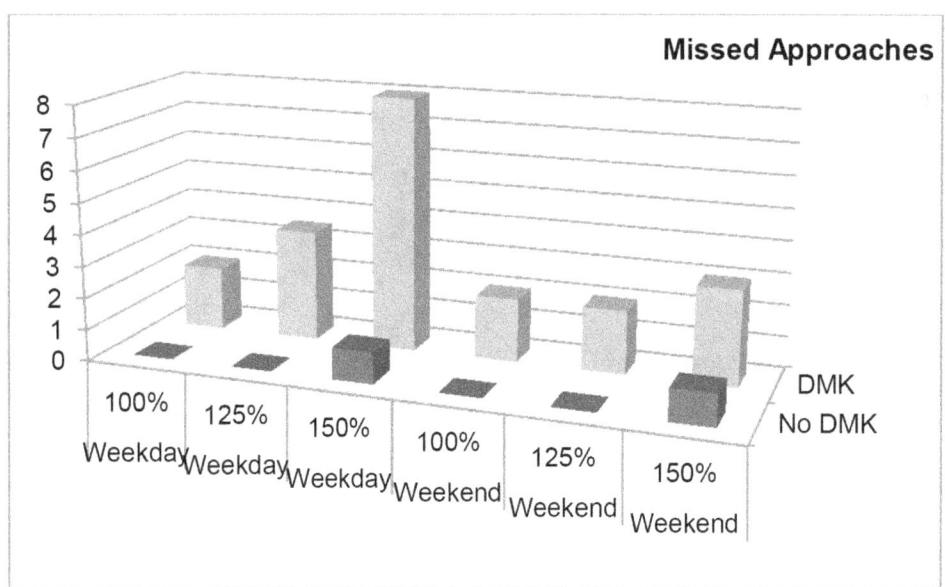

Figure 41 - DMK Minor Separation Errors and Missed Approaches

Figure 41 shows that minor separation errors (mostly slight infringements of horizontal separation) are low, except when working 150% samples with DUMMKOPF – in which the approach controller lost control on several occasions. This is confirmed by the frequency of missed approaches.

Figure 42 summarises the effects of **DOW** the day of week and **TOD** the time of day), with the effect of **TRA** (the level of traffic) for Hentzau and Zenda TMAs, on the mean arrival and departure delays as recorded in tables 38 and 39, for weekday and weekend traffic..

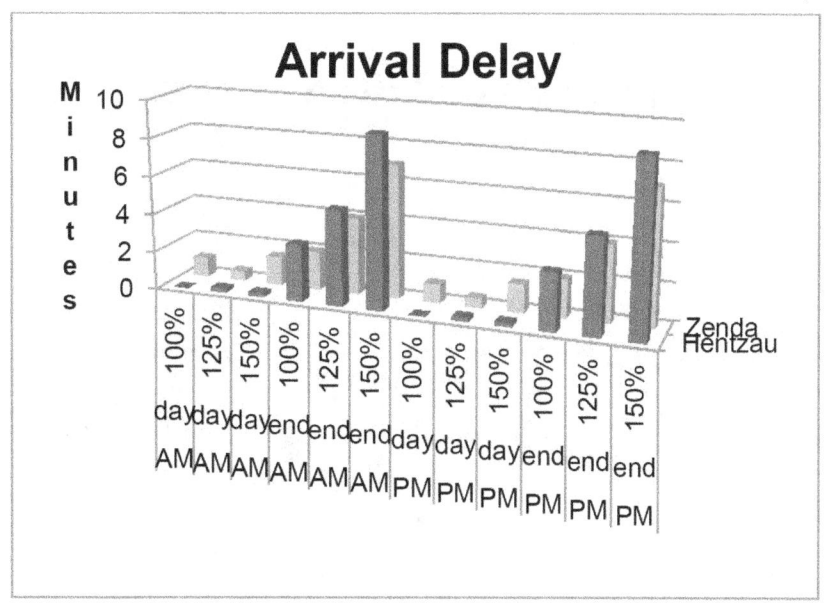

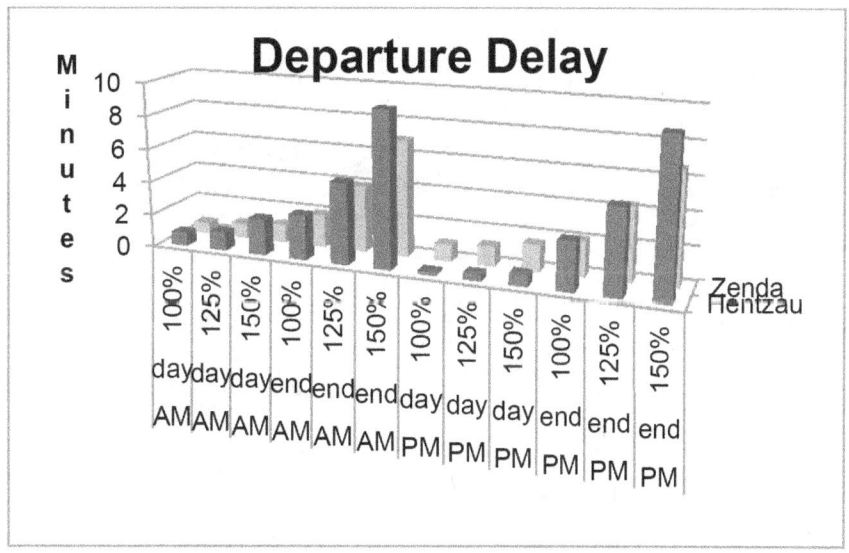

Figure 42 - Hentzau / Zenda Delays

Figure 42 shows that delays are negligible during weekdays, but increase sharply with 150% traffic samples. The controllers maintain an even distribution of delay, although traffic is predominantly arrivals in the morning and departures in the afternoon. DUMMKOPF had no effect on these two TMAs.

Figure 43 summarises the effects of **DOW** the day of week and **TOD** the time of day), with the effect of **TRA** (THE LEVEL OF TRAFFIC) for Hentzau and Zenda TMAs, on the mean numbers of minor separation errors and missed approaches as recorded in Figures 38 and 39, for weekday and weekend traffic..

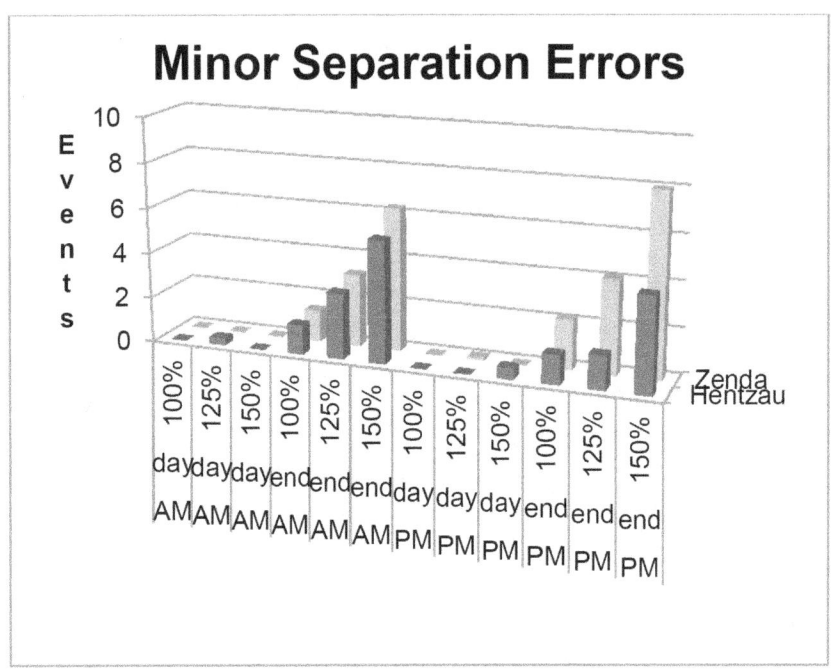

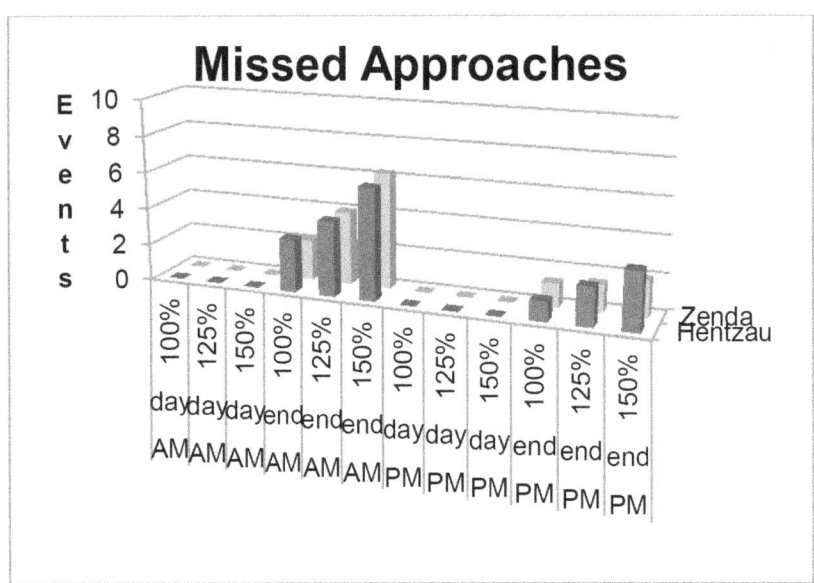

Figure 43 - Hentzau / Zenda Separation/ Missed Approaches

Figure 43 shows that separation errors and missed approaches are negligible during weekday traffic, but that separation errors increase in line with traffic, and that missed approaches increase sharply for weekend morning traffic for both TMAs.

TMA - Factors - Working Position Level

The first three TMA ANOVAR factors shown in Figure 44 are the factors between exercises, as shown in the exercise title. In addition, there will be separate files for the planner(s) and executive(s) within the exercises, giving an additional factor - **WORKING POSITION (WkP)** with a value of 1 for the TMA controller and 2 for the Approach controller. Here, as at the TMA level, the three TMAs are analysed separately.

Figure 39 shows the 13 potentially significant factors and interactions at the working position level.

A full report would require a tabulation for 18 variates x 6 anovars (weekday/weekend x TMA) x 13 factors and interactions - 1404 tabulations in all. These can be supplied on request but in the interests of simplicity, only 'significant' interactions are reported, and illustrated graphically or by tabulation.

The sensitivity of the ANOVAR depends on the number of residual degrees of freedom. Non-significant interactions are therefore folded into the residual, the least significant first, until either there are no more non-significant interactions, or only the main factors are left, whether they are significant or not.

In the following tables, three stars represent a difference significant at the 0.1% level. Two stars represent a difference significant at the 1% level. One star represents a difference significant at the 5% level. A blank cell indicates no significant difference. To economise on

Factor	Degrees of Freedom
Dmk	1
ToD	1
Tra	2
WkP	1
Dmk x ToD	1
Dmk x Tra	2
Dmk x WkP	1
ToD x Tra	2
ToD x WkP	1
Tra x WkP	2
Dmk x ToD x Tra	2
Dmk x ToD x WkP	1
ToD x Tra x WkP	2
Residual	4
TOTAL	23

Figure 44 TMA ANOVAR Working Position

space, interactions that are not significant for any variate – a completely blank line - are omitted. Equally, variates that are not affected significantly by any factor are omitted.

Because there are so many variates involved, the factor/ variate tabulation has been split into two parts – the first containing 'communications' measures, and the second 'subjective' (ISA AND NASA TLX)

StrelsauTMA								
Factor	NFr	TFr	NIn	Tin	NHo	THo	NOr	TOr
Dmk		*		*		*	*	*
Tra	*	*	**	*	**	*	**	*
WkP	*	**	**	**	**	**	**	**
Dmk x WkP	*	*	*	*	*	*	*	**

Hentzau TMA								
Factor	NFr	TFr	NIn	Tin	NHo	THo	NOr	TOr
Tra	**	**	**	**	**	**	**	
WkP	**	**	*	*	*	*	*	*
ToD x Tra	*		*		*			*
ToD x WkP	*	*		*		*	*	*
Tra x WkP	*	*	*	*	*	*	*	*

Zenda TMA								
Factor	NFr	TFr	NIn	Tin	NHo	THo	NOr	TOr
ToD		*		*	*	*		
Tra	**	**	**	**	**	**	**	
WkP	**	**	*	*	*	*	*	*
ToD x Tra	*		*		*			*
ToD x WkP		*		*	*	*	*	
Tra x WkP	*		*		*	*		

Figure 45 - Communications Factors / Work Position Weekdays

Figure 45 summarises the analyses (figure 44) for weekday traffic of the 10 working position 'communications' level variates, (figure 16) giving the significance of remaining differences for the three TMAs separately.

Control Room Simulation

Strelsau TMA								
Factor	NFr	TFr	NIn	Tin	NHo	THo	NOr	TOr
Dmk		*		*		*		
Tra	*	**	**	**	**	**	**	
WkP	**	**	*	**	*	**	**	*
Dmk x Tra	*		*		*			*
Dmk x WkP	*	*		*	*		*	*

Hentzau TMA								
Factor	NFr	TFr	NIn	Tin	NHo	THo	NOr	TOr
ToD		*		*		*		*
Tra	**	**	**	**	**	**	**	
WkP	**	**	*	*	*	*		
ToD x Tra	*		*		*			*
ToD x WkP	*	**	*	**	*	*	*	*

Zenda TMA								
Factor	NFr	TFr	NIn	Tin	NHo	THo	NOr	TOr
ToD		*		*		*		*
Tra	*		*	*	**	**	**	
WkP	**	**	*	*	*	*	*	*
ToD x Tra	*		*		*			*
ToD x WkP	**	*	*	*	*	**	*	*
Tra x WkP	*	*	*	*		*		*

Figure 46 - Communications Factors / Work Position Weekends

Figure 46 summarises the analyses (figure 44) for weekend traffic of the 10 working position 'communications' level variates, (figure 16) giving the significance of remaining differences for the three TMA separately.

A2-53

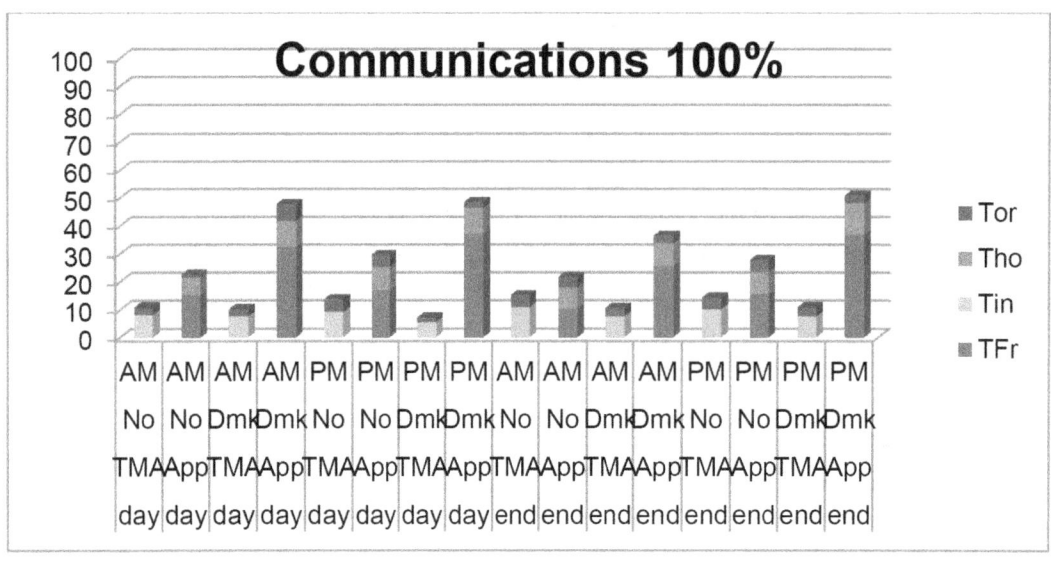
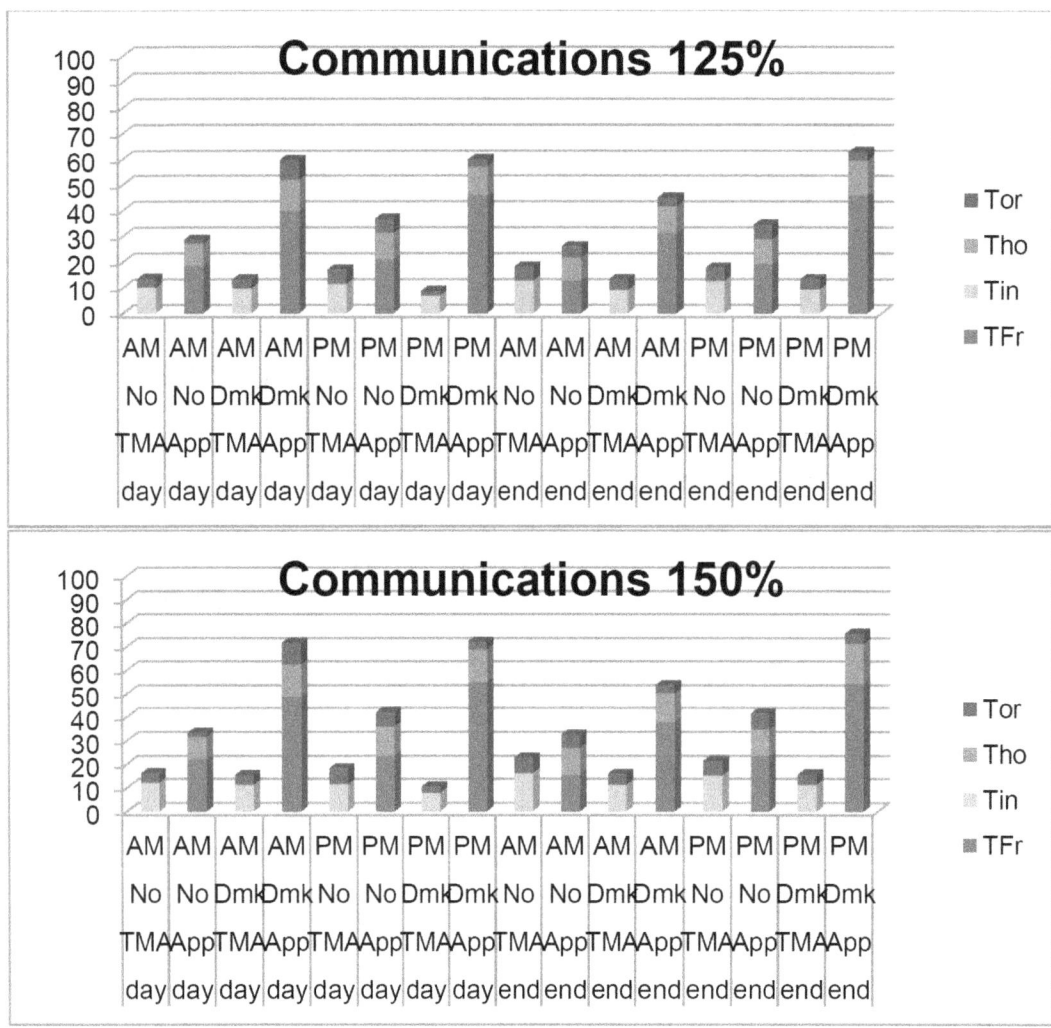

Figure 47 - Strelsau Communications / Dummkopf

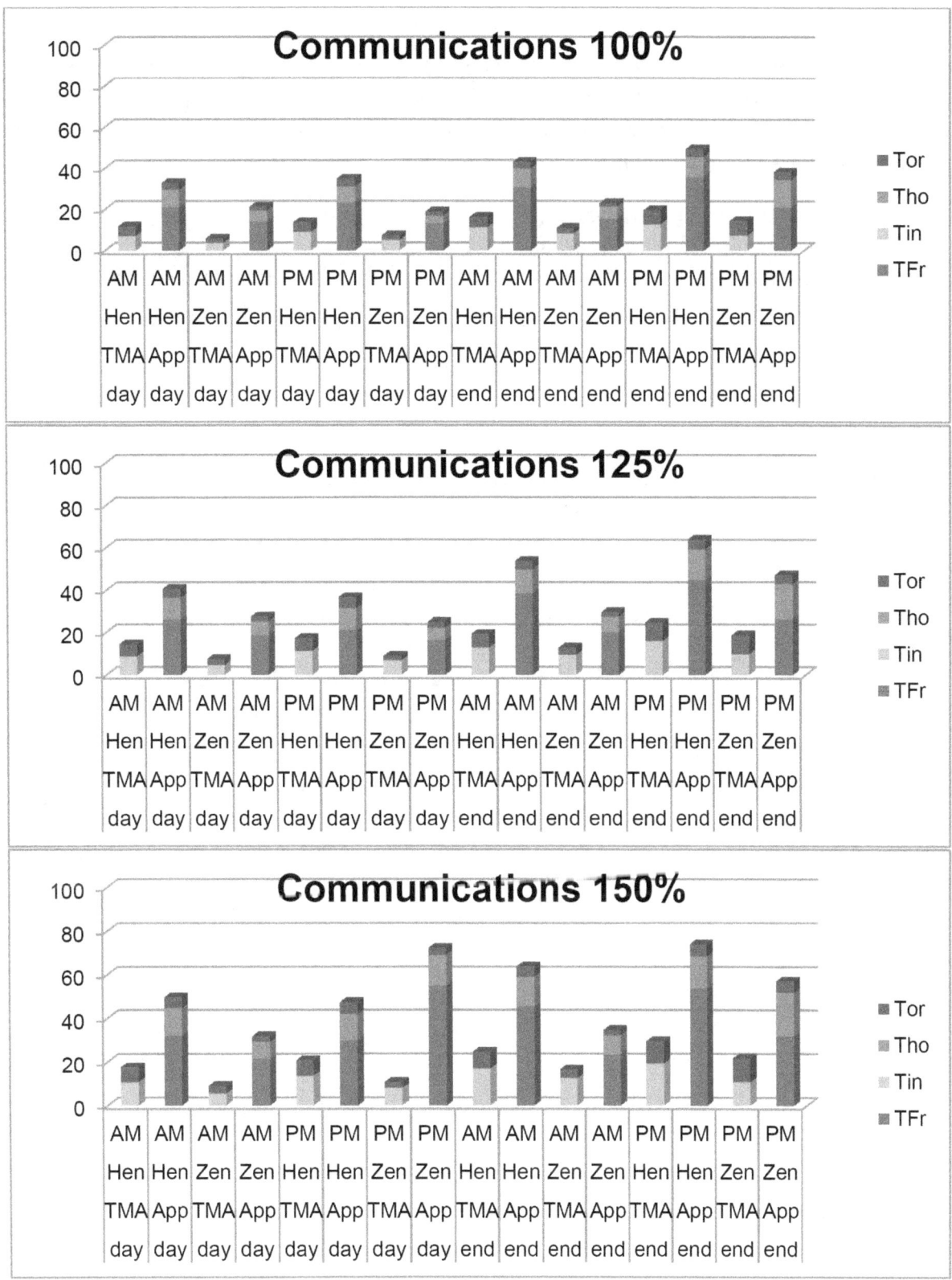

Figure 48 - Hentzau / Zenda Communications

Figure 47 summarises the analyses (figures 45 and 46) of the 4 'percentage time' communications variates for working position for Strelsau TMA , showing the difference due to **Dmk**, for different traffic levels. The numbers of communications echo these percentage differences, and are therefore not included -, to simplify the presentation.

Figure 48 summarises the analyses (Figures 45 and 46) of the 4 'percentage time' communications variates for Working Position for Hentzau and Zenda TMAs , which are not affected by **Dmk** for different traffic levels. The numbers of communications echo these percentage differences, and are therefore not included - to simplify the presentation

Control Room Simulation

Figure 49 summarises the analyses (Figure 44) for weekday traffic of the 8 Working Position 'subjective' variates, (Figure 17) giving the significance of remaining differences for the three TMAs separately.

Strelsau TMA							
Factor	VISA	TISA	TLMe	TLPh	TLTE	TLEf	TLFr
Dmk		*		*		*	
Tra	*	**	**	**	**	**	**
WkP	**	**	*	**	*	**	**
Dmk x Tra	*		*		*		
Dmk x WkP	*	*		*	*		*

Hentzau TMA							
Factor	VISA	TISA	TLMe	TLPh	TLTe	TLEf	TLFr
ToD			*		*		*
Tra	*	*		*	*	**	**
WkP	**	*	**	*	*	*	*
Tra x WkP	*		*	*	*		*

Zenda TMA							
Factor	VISA	TISA	TLMe	TLPh	TLTe	TLEf	TLFr
ToD			*		*		*
Tra	*	*		*	*	**	**
WkP	**	*	**	*	*	*	*
ToD x WkP	**		*	*	*	*	**
Tra x WkP	*		*	*	*		*

Figure 49 - TMA Subjective / Work Position Weekdays

Figure 50 summarises the analyses (Figure 44) for weekend traffic of the 8 working position 'subjective' variates, (Figure 17) giving the significance of remaining differences, for the three TMAs separately.

Strelsau TMA							
Factor	VISA	TISA	TLMe	TLPh	TLTE	TLEf	TLFr
Dmk				*		*	
Tra	*		**	**	**	**	**
WkP	**	*	*	**	*	**	**
Dmk x Tra	*		*		*		
Dmk x WkP	*	*		*	*		*

Hentzau TMA							
Factor	VISA	TISA	TLMe	TLPh	TLTe	TLEf	TLFr
Tra	*	*		*	*	**	**
WkP	**	*	**	*	*	*	*
ToD x Tra	*			*		*	
ToD x WkP	**		*	*	*	*	**
Tra x WkP	*		*	*	*		*

Zenda TMA							
Factor	VISA	TISA	TLMe	TLPh	TLTe	TLEf	TLFr
ToD			*		*		*
Tra	*	*		*	*	**	**
WkP	**	*	**	*	*	*	*
ToD x WkP	**		*	*	*	*	**
Tra x WkP	*		*	*	*		*

Figure 50 - TMA Subjective / Work Position Weekends

Control Room Simulation

Figure 51 illustrates the mean value of ISA ((Instantaneous Self-Assessment) for Strelsau TMA, by Working Position, presence or absence of DUMMKOPF and traffic level for each Time of Day and Day of Week.

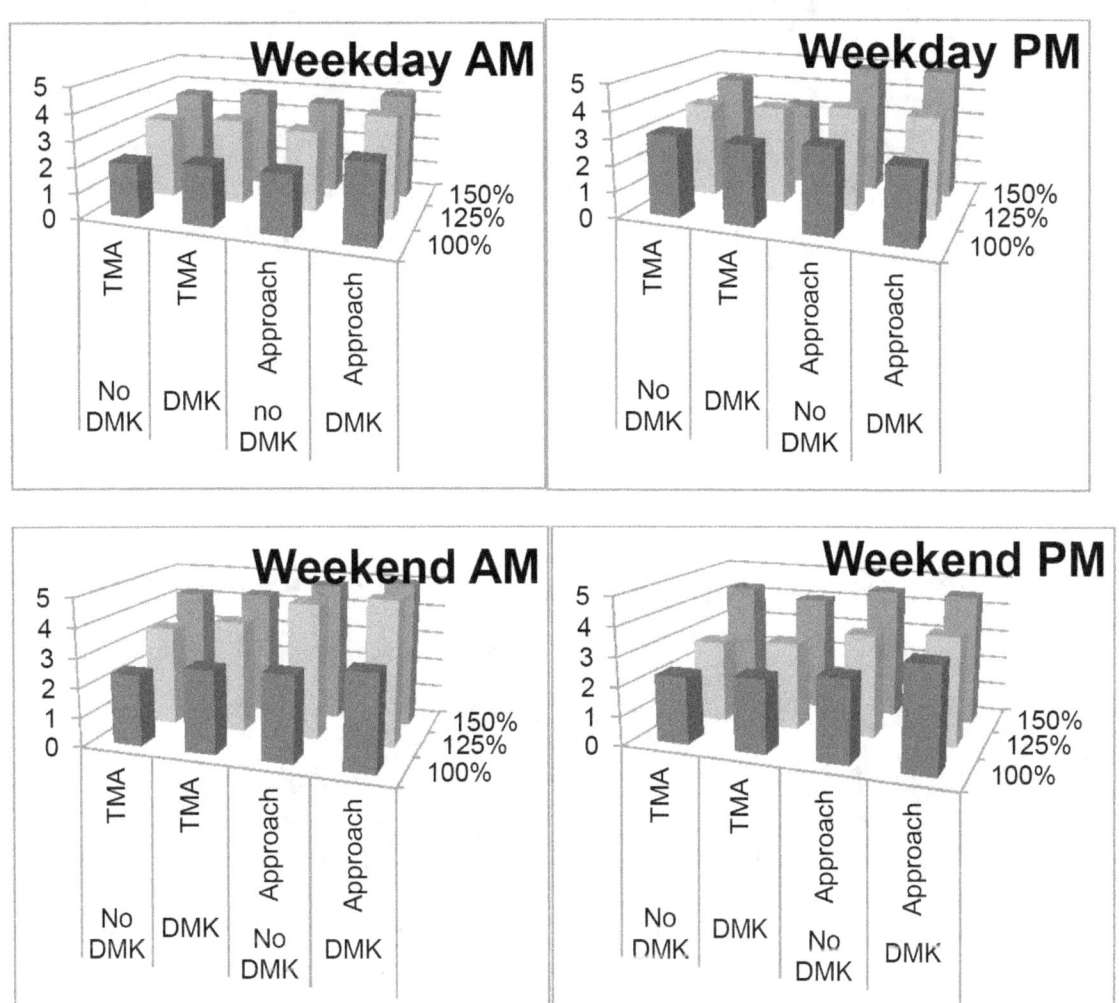

Figure 51 - Strelsau Work Position / DUMMKOPF ISA

Figure 52 illustrates the mean value of ISA (Instantaneous Self-Assessment) for Hentzau and Zenda TMAs, by working position, and traffic level for each time of day and day of week.

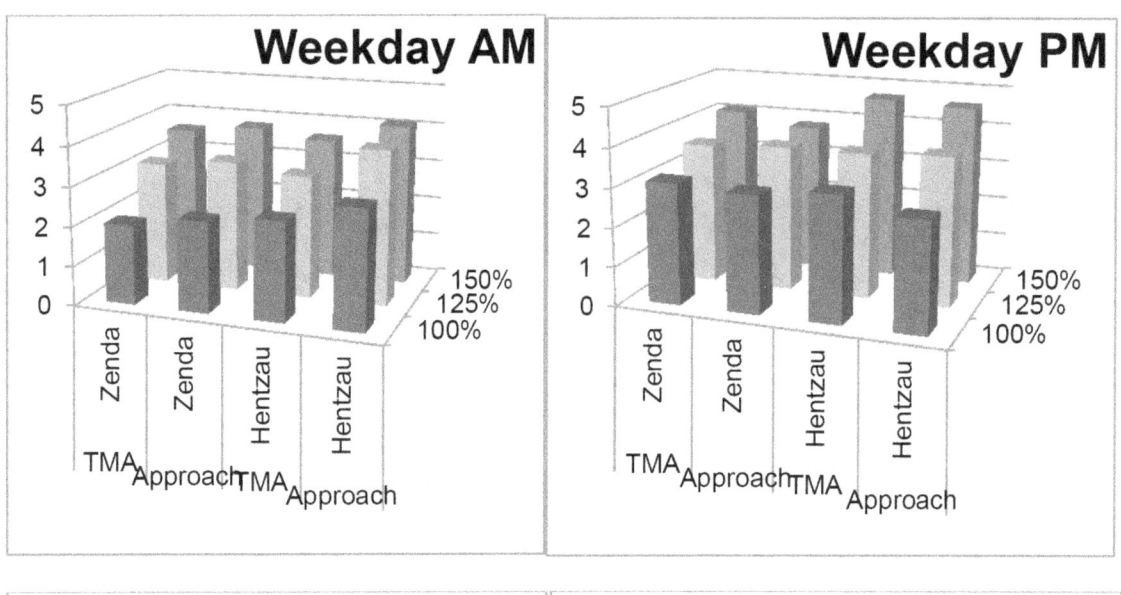

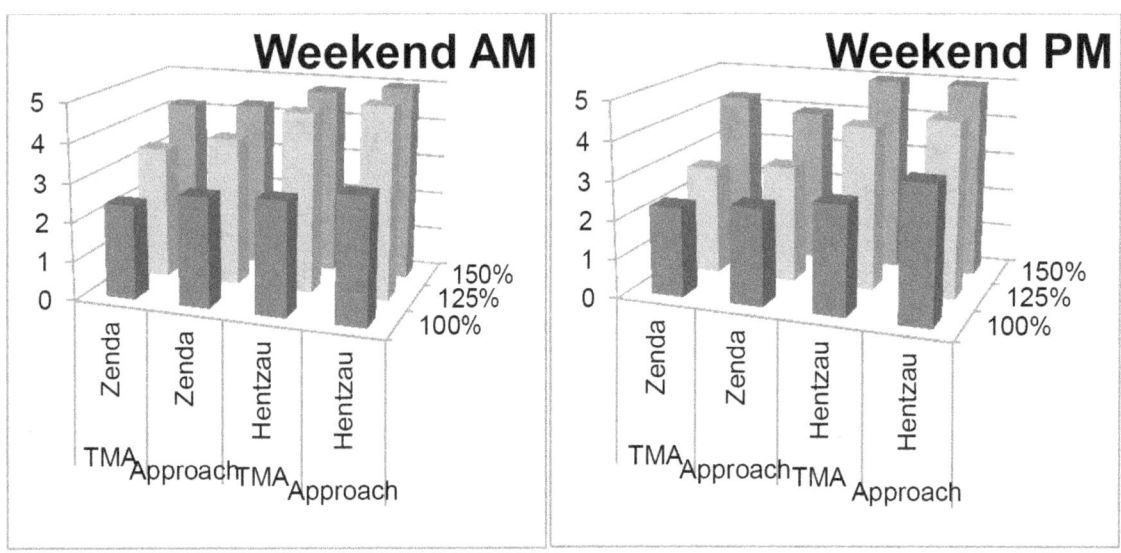

Figure 52 - Hentzau / Zenda Work Position ISA

Figure 53 illustrates the mean value of the six scales of the NASA TLX (Task Load Index) for Strelsau TMA by Working Position and presence or absence of DUMMKOPF for each Traffic Level (Time of Day and Day of Week do not differ significantly).

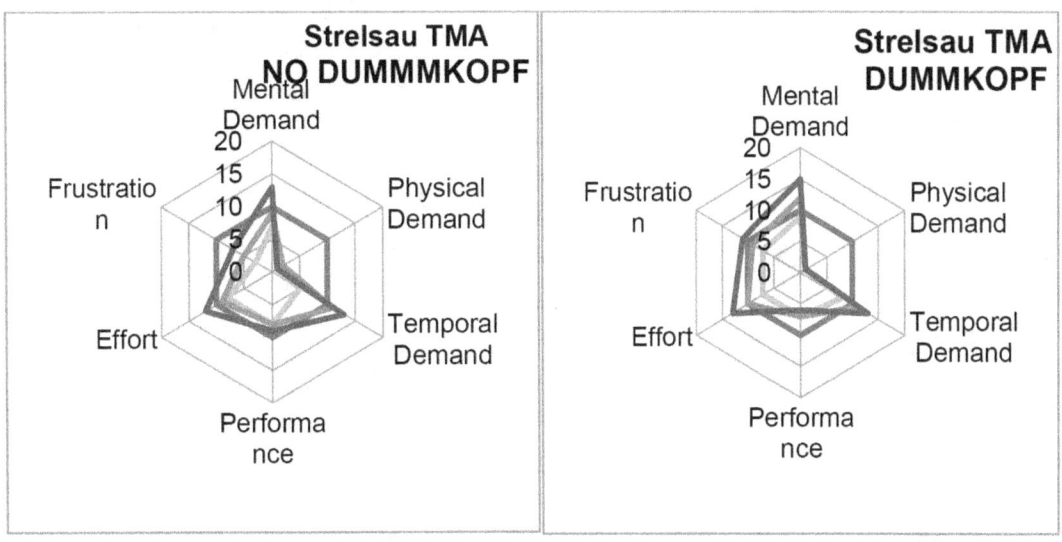

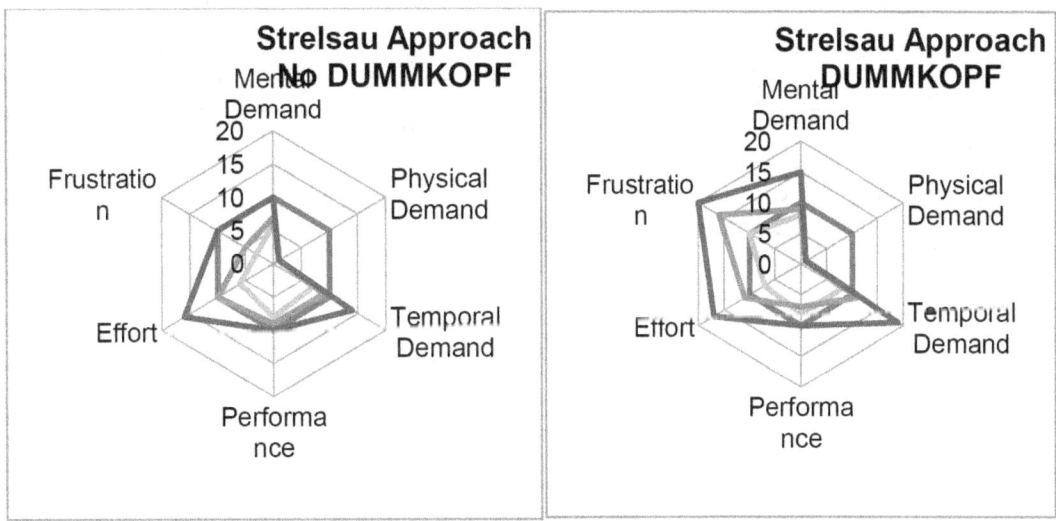

Figure 53 - Strelsau NASA TLX

Figure 54 illustrates the mean value of the six scales of the NASA TLX (Task Load Index) for Hentzau and Zenda TMAs by Working Position for each Traffic Level (Presence or absence of DUMMKOPF, Time of Day and Day of Week do not differ significantly).

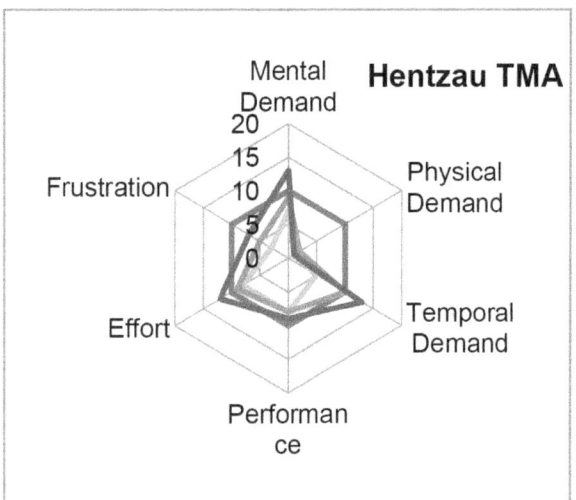
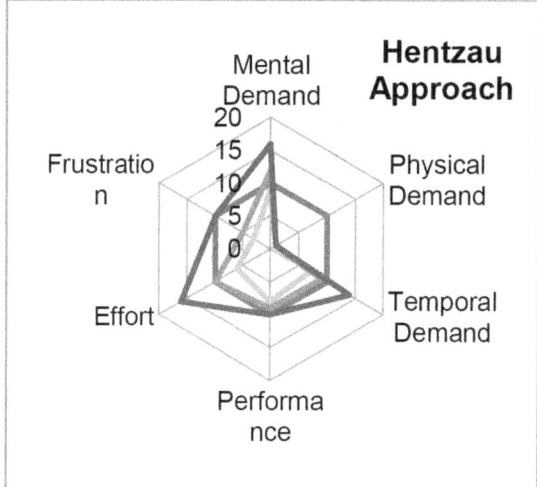
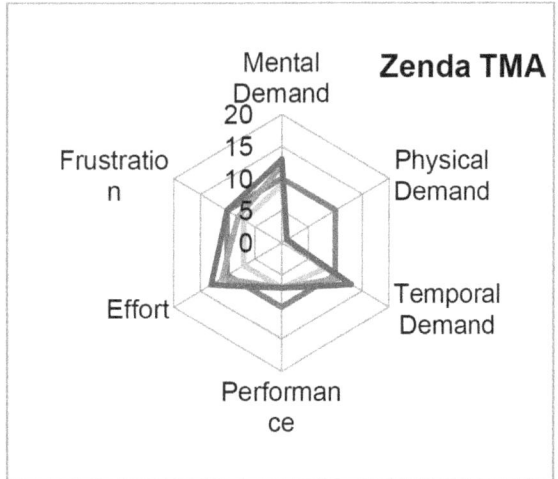
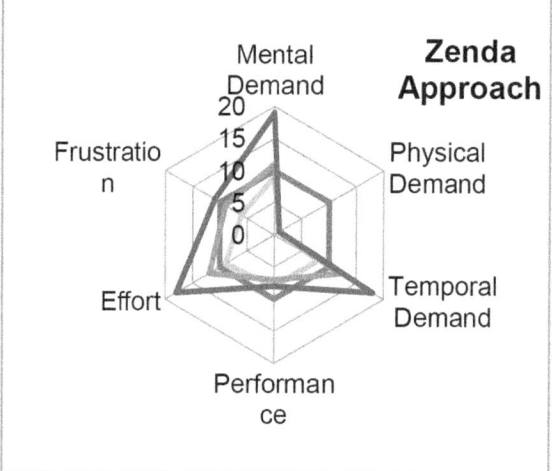

Figure 54 - Hentzau / Zenda NASA TLX

TMA - Discussion

The purpose of the inclusion of TMAs in this simulation was to evaluate DUMMKOPF (Digital Universal Monitoring and Mediation Kontextual Operational Planning Function) at Strelsau International Airport.

It is clear that DUMMKOPF, in its present form, is not a significantly useful tool for routine ATC operation. Because various constraints have not been taken into account, and because the integration of the DUMMKOPF software into the EREHWONTROL Experimental Centre simulator occasioned significant problems, the controllers were never able to rely on DUMMKOPF recommendations. One controller pointed out that when DUMMKOPF was in use, the approach controller had to follow what DUMMKOPF was doing, understand why it was doing it, check that the proposals were safe, and maintain an alternative plan in case DUMMKOPF malfunctioned. This involved at least three times as much workload as working without DUMMKOPF, with little appreciable advantage to traffic. Frustration levels were considerable.

Nevertheless, controllers accepted that some form of aid was necessary for approach control, provided that it had proven reliability.

Unexpectedly, it became clear that the major traffic problem was not 'en-route', nor at Strelsau International, but at Hentzau and Zenda Airports. Weekend morning arrivals and evening departures placed considerably greater strain on these airports than occurred during the week anywhere or at Strelsau International at weekends.

This is, of course, assuming that the current proportions of traffic do not vary as traffic grows. A considerable proportion of the traffic at Hentzau Airport consists of training flights by the Ruritanian Civil and Commercial Aviation Training Institute. Re-location of this traffic, or even simply avoiding peak times, would considerably relieve this problem.

As with the en-route sectors, the equipment and procedures for all TMAs in Ruritania are generally obsolescent, and a program of modernisation is needed.

Chapter 7 – Conclusions

1. Will the proposed division of Ruritanian Airspace provide increased capacity during normal week-day operation?

   Yes, the proposed division into Ruritania East and Ruritania West will provide about 20% increase in week-day capacity. Given that the present system can handle up to 150% of current traffic, the two sector organisation should handle up to 170% of Current Weekday traffic.

2. Will the proposed division of Ruritanian airspace provide increased capacity during week-end operation?

   Yes, the proposed division into Ruritania East and Ruritania West will provide about 20% increase in week-end capacity. Given that the present system can handle up to 140% of current traffic, the two sector organisation should handle up to 160% of Current Weekday traffic.

3. Is the Digital Universal Monitoring and Mediation Kontextual Operational Planning Function (DUMMKOPF) an acceptable function for Ruritanian Airspace?

   No.
   There are certain situations where the DUMMKOPF function 'loses' aircraft, leading to potential conflicts.. Where an aid is not trusted, the controllers are obliged to observe its recommendations, evaluate them for safety, and maintain an on-going picture of what action to take in case of an error or malfunction. This considerably increases workload, time stress and frustration. In addition, there were some problems in integration with the EEC simulator, which are likely to recur in its integration into the Ruritanian Airspace Control system.

   Additional Conclusions

4. About 25% of the arrivals and departures at Hentzau, on weekdays and weekends, consists of training flights operated by the Ruritanian Civil and Commercial Aviation Training Institute. These flights should be halted during peak periods. In the long-term interests of safety and economy. RCCATI flight operations should be transferred to a less busy airport –possibly Martburg Civil. Aircraft maintenance, theoretical and simulation training can be maintained at Hentzau.

5. The general level of equipment and procedures in Ruritanian ATC is obsolescent. A more modern system is required.

Acknowledgements

The Erewhontrol Experimental Centre gratefully acknowledges the significant contributions made to the successful completion of this simulation by:

Dhr. R. Hentzau. (Ruritania. Control),
Dfr. A de Maupin (Ruritania. Control),
Dhr. R.S. Loch (Ruritania. Control),
Dhr. Professor V. Frankenstein (University of Strelsau.),
Dhr. Docent I. Igorovitch (University of Strelsau.),

and by the controllers and assistants from Ruritania. Control, Strelsau. International, Hentzau. and Zenda. Airports.

Appendix 3 – Conference Paper

The following conference paper is compiled in accordance with the specification for a recent annual conference of a small learned society in the United Kingdom. Since the society in question no longer produces printed versions of conference proceedings, some restrictions on, for example, the use of colour illustrations and the maximum number of pages have been relaxed.

Unfortunately, colour can not be used in this book.

Ruritania Air Traffic Control Simulation

E. Bennet[3]

ABSTRACT

Ruritanian airspace is approaching saturation. This simulation investigated the possible increase in capacity from dividing the airspace into two sectors. It also investigated the value of a proposed semi-automatic arrival sequencing aid. Traffic samples corresponding to morning and evening 95%ile traffic were used, at a density of 100%, 125%, 150% and 175% of current traffic. The present single-sector organisation could cope with up to 150% of current traffic on weekdays, and 140% of current traffic at weekends, while the two- sector organisation could cope with up to 20% more traffic. Neither organisation could handle 175% of existing traffic. Unexpectedly, the saturation point was Hentzau TMA, not Strelsau International. Some relief is available from restricting training flights at peak periods. The proposed sequencing aid did not improve performance, and controllers felt that it increased their workload. There were significantly more delays and separation infringements when it was in use. Considerable difficulty was experienced in integrating the sequencing aid into the Erehwontrol Experimental Centre simulator.

KEYWORDS

Air Traffic Control, Sectorisation, Arrival sequencing aid

Introduction

Ruritania is a nation state situated in the western part of the continent of Erehwon, bordered on the north by Kennaquhair, on the east by Datong, on the south by Nephelocccygia and on the west by Atlantis. The main airport is Strelsau. International. Ruritani is a member of Erehwontrol, which coordinates the Air Traffic Control services of the nations of Western Erehwon. In view of increasing international traffic, the Ruritanian Ministry of Aviation is considering splitting Ruritanian airspace into two sectors instead of the one existing (Figure 1). They are concerned that the additional co-ordination workload may balance out the reduction in traffic under control.

In addition, the Ministry is interested in evaluating a computer-based sequencing aid for departures and arrivals at Strelsau. International, avoiding conflicts with local traffic from Hentzau and Zenda. airports. The procedure, developed by the well-known Artificial Intelligence expert, Dr V. Frankenstein, needed to be evaluated.

---

[3] EREHWONCONTROL Experimental Centre, Utopia

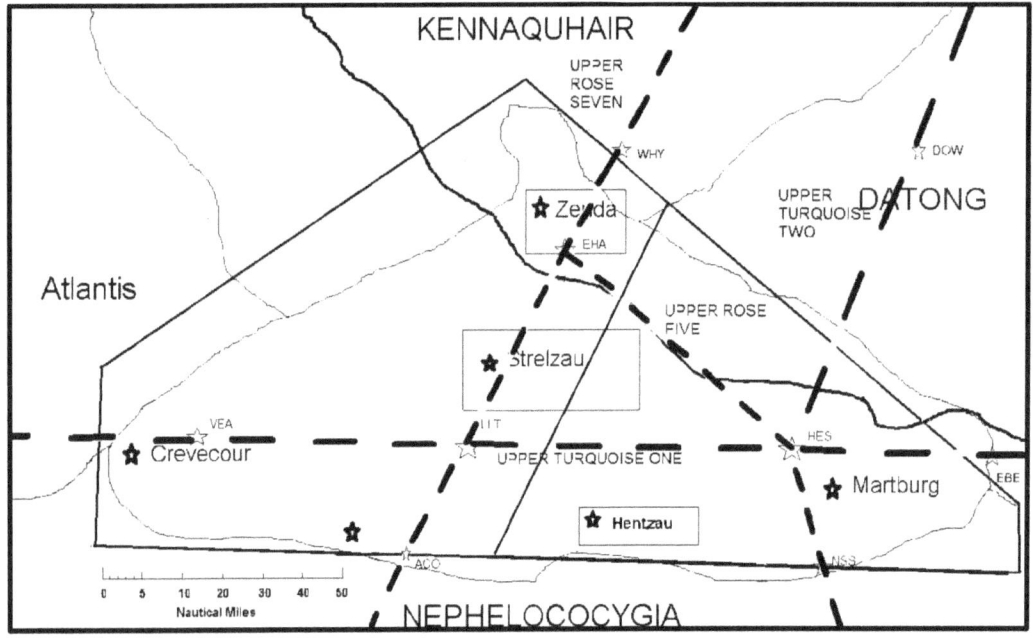

Figure 1 – Proposed Two- sector Ruritania Airspace

Method

A real-time simulation was carried out at the Erehwon Experimental Centre in October 2020, comparing one and two sector organisations, with weekday and weekend traffic at 100%, 125%, 150% and 175% of current 95-percentile traffic. Strelsau International TMA was simulated, with and without the Digital Universal Monitoring and Mediation Kontextual Operational Planning Function (DUMMKOPF). Hentzau and Zenda TMAs were also simulated. Figure 2 shows the simulation Control room, (When one sector was simulated, the Ruritania west sector suite was used, to provide better viewing by the supervisors.) Two teams of controllers took part, simulating one or two sector teams, with three TMA teams, each of two controllers – eight controllers for the one sector or ten controllers for the two-sector organisation

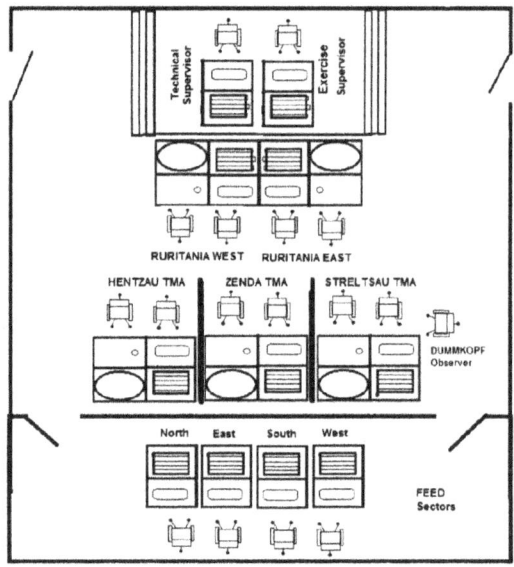

Figure 2 – Ruritania I Simulation Control Room

ISA (Instantaneous Self-Assessment) was employed at three minute intervals throughout each 90 minute run, (The mean ISA from 15 – 75 min was used for analysis, to avoid initial and final effects) and the NASA-TLX questionnaire was applied at the end of each run, with a run-specific and general questionnaire. Normal measures of traffic and communications loading, and of separation infringements, missed approaches and delayed departures or landings were applied.

Six initial training and familiarisation sessions were carried out before the measured runs. Blocks of eight 90-minute runs were run with increasing traffic load,. Each block balanced Morning/Afternoon, Weekday/Weekend and One-sector/Two-sector for each team. DUMMKOPF was also applied to Strelsau International in a balanced plan. The initial 100% block was badly affected by problems arising from the integration of DUMMKOPF, to the extent that it could not be safely included in the final analysis. Blocks of 125% and 150% traffic were simulated successfully, but it was clear that 175% of current traffic was not workable and most of these exercises were terminated well before completion. This block was therefore also excluded from the analysis. The final repeat of the 100% traffic was satisfactory, and was therefore included in the analysis.

Results

Sectorisation

Four factors were examined in the analysis of sectorisation: Number of Sectors, East versus West, Morning vs Afternoon and Traffic level. Weekday and Weekend traffic were examined separately. Analyses were carried out at sector level (5 objective variates) and working position level (10 objective and 8 subjective variates.) Although the 46 analyses of variance and the 692 tables of mean values are available from the author, only the most interesting results can be given here

As was expected, the Single Sector layout involved significantly greater individual workloads than the Two Sector Layout. However, more detailed examination shows some interesting and unexpected features.

Given the existence of three major airports, it was not possible to divide the single Sector layout into two equal parts. Consequently, Ruritania West had a workload of about 70% of that of the Single sector, and Ruritania East had a relatively light workload of about 40% of the Single Sector. (The extra 10% is generated by the interface between Ruritania East and Ruritania West.)

The Single Sector layout can cope well with the current (100%) 95%ile load, and with difficulty with 125%, but cannot safely handle 150%. These levels correspond to the anticipated traffic in four and seven years.

The Two Sector layout would handle up to an estimated 170% traffic, corresponding to the anticipated traffic in ten years' time.

The traffic we are here considering is the 95% ile, occurring about 18 days in the year. Temporary traffic flow restrictions on these days may be acceptable for a few years.

Equally, these projections assume no significant change in the mixture of traffic in future years. Particularly at weekends, some traffic might be diverted from Hentzau, and to a lesser extent from Zenda. to Strelsau. international which is less loaded at weekends.

Additionally, it appears that a significant proportion of Hentzau traffic originates with the Ruritanian Civil and Commercial Aviation Training Institute, flying short training flights. If these could be moved elsewhere or simply moved away from peak times, a substantial reduction of peak traffic can be achieved.

Finally, It should be noted that the equipment and procedures at Ruritania Control are generally rather 'traditional' and that updating of equipment and procedures would be welcomed by the controllers, and provide increased capacity and safety.

TMAs

The purpose of the inclusion of TMAs in this simulation was to evaluate DUMMKOPF (Digital Universal Monitoring and Mediation Kontextual Operational Planning Function) at Strelsau International Airport. Zenda and Hentzau were included for completeness.

Two factors were examined in the analysis of TMAs: morning vs afternoon and traffic level.. The presence or absence of DUMMKOPF was a third factor for Strelsau TMA only. Weekday and weekend traffic were examined separately for each TMA analyses were carried out at TMA level (10 objective variates) and working position level (10 objective and 8 subjective variates.) Although the 168 analyses of variance and the 1160 tables of mean values are available from the author, only the most interesting results can be given here.

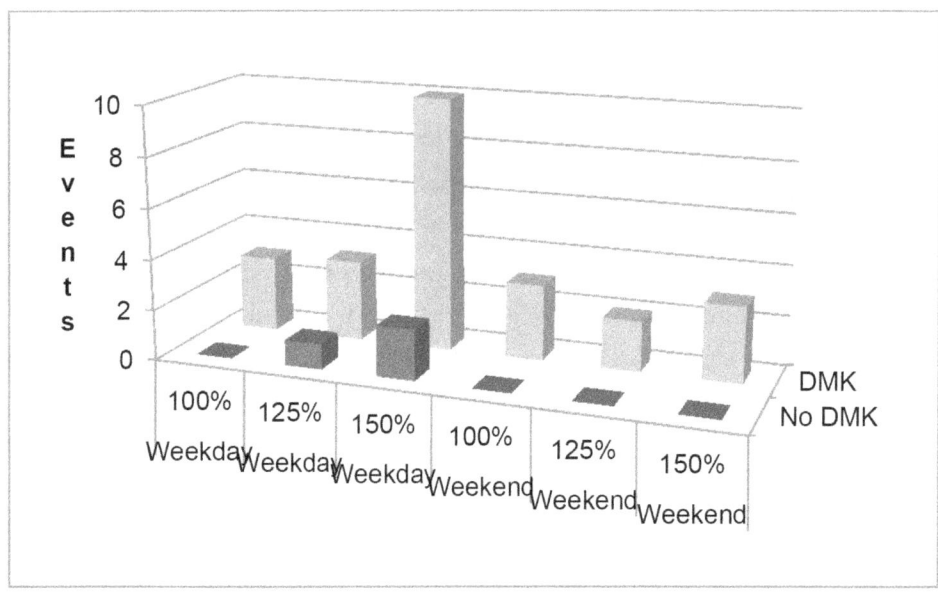

Figure 3 - Minor Separation Errors

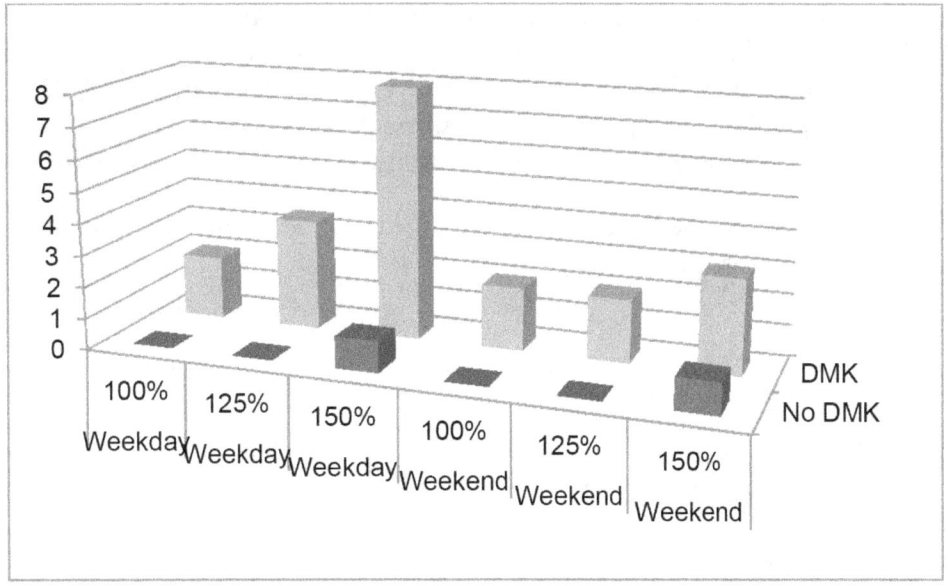

Figure 4 - Missed Approaches

Figure 3 shows the incidence of minor separation errors (>4 nm minimum separation) in Strelsau TMA, with and without DUMMKOPF. (There were too few major errors for analysis) It shows that there were significantly more errors when DUMMKOPF was in use, particularly with heavy traffic. Figure 4 confirms this observation for missed approaches. It appears that the approach controllers 'lost the picture' repeatedly when they were constrained by DUMMKOPF.

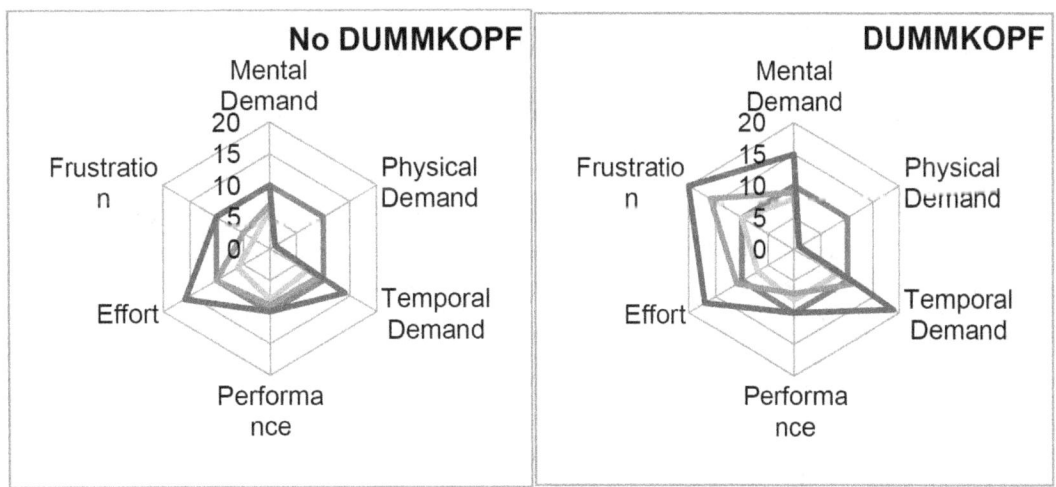

Figure 5 Strelsau Approach - NASA-TLX Component Scores

Figure 5 shows that that while Strelsau approach controllers felt that they were fully occupied with 150% traffic without DUMMKOPF, they felt overloaded both for temporal demand and frustration when obliged to use DUMMKOPF.

Because various constraints have not been taken into account, and because the integration of the DUMMKOPF software into the EREHWONTROL Experimental Centre simulator occasioned significant problems, the controllers were never able to rely on DUMMKOPF recommendations. One controller pointed out that when DUMMKOPF was in use, the approach controller had to follow what DUMMKOPF was doing, understand why it was doing it, check that the proposals were safe, and maintain an alternative plan in case DUMMKOPF malfunctioned. This involved at least three times as much workload as working without DUMMKOPF, with little appreciable advantage to traffic. Frustration levels were considerable.

Nevertheless, controllers accepted that some form of aid was necessary for approach control, provided that had proven reliability.

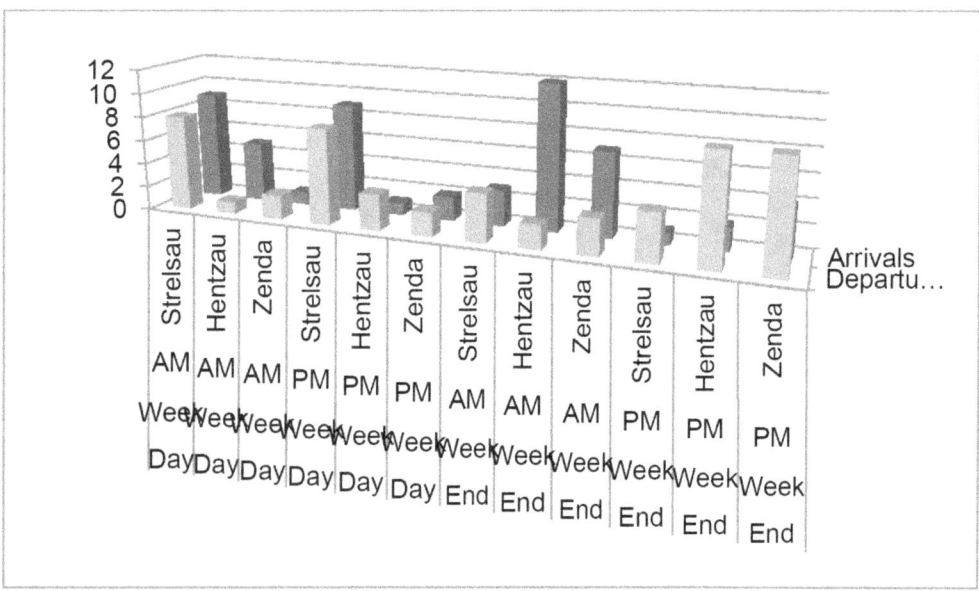

Figure 5 - Arrivals / Departures AM/PM Weekday/Weekend by TMA

Figure 5 shows the current 95% le peak hourly rates for Morning and Evening Peaks, for Weekdays and Weekends

Unexpectedly, it became clear that the major traffic problem was not 'en-route', nor at Strelsau International, but at Zenda and particularly Hentzau Airports. Weekend morning arrivals and evening departures placed considerably greater strain on these airports than occurred during the week anywhere or at Strelsau International at weekends.

Discussion

The traffic we are here considering is the 95th percentile, occurring about 18 days in the year. Temporary traffic flow restrictions on these days may be acceptable for a few years.

The extrapolations to 125% and 150% assume no significant change in the mixture of traffic in future years. Particularly at weekends, some traffic might be diverted from Hentzau, and to a lesser extent from Zenda. to Strelsau. international which is less loaded at weekends.

Additionally, it appears that a significant proportion of Hentzau traffic originates with the Ruritanian Civil and Commercial Aviation Training Institute, flying short training flights. If these could be moved elsewhere or simply moved away from peak times, a substantial reduction of peak traffic can be achieved.

Finally, It should be noted that the equipment and procedures at Ruritania Control are generally rather 'traditional' and that updating of equipment and procedures would be welcomed by the controllers, and provide increased capacity and safety. This applies to en-route sectors and TMAs alike.

Conclusions

The existing one sector organisation is capable of handling 150% of current peak traffic on weekdays, and 140% at weekends.

The two sector organisation can handle 20% more traffic in either case.

DUNNKOPF is not yet in a suitable state for incorporation in the Ruritanian ATCV system, although it has promising aspects.

The most stressed part of the existing system is Hentzau Approach, particularly on weekend mornings, where there is a large number of arrivals.

Some relief may be obtained by restricting training flights during peak periods.

Acknowledgements

The Erehwontrol Experimental Centre acknowledges the whole-hearted cooperation of the controllers from Ruritania Control, Strelsau, Hentzau and Zenda TMAs and of support staff from the Ruritanian Ministry of Aviation. The outstanding contribution of Dhr. Rupert Hentzau will be long remembered.

Appendix 4 – Verbal Presentation

Slide 1

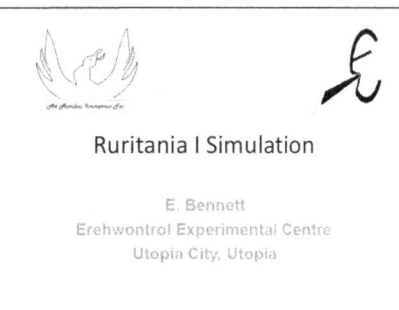

Ruritania I Simulation

E. Bennett
Erehwontrol Experimental Centre
Utopia City, Utopia

Good Morning, Ladies and Gentlemen, I am Squadron Leader(Retired) Elizabeth Bennett
I am a simulation leader at The Erehwontrol Experimental Centre
The Ruritania I simulation took place from October 12 to October 30 2020

Slide 2

Simulation Aims

- Will the proposed division of Ruritanian airspace provide increased capacity during normal week-day operation?
- Will the proposed division of Ruritanian airspace provide increased capacity during week-end operation?
- Is the Digital Universal Monitoring and Mediation Kontextual Operational Planning Function (DUMMKOPF) an acceptable function for Ruritanian Airspace?

The Ruritanian Air Ministry was concerned about the ability of Ruritania Control to handle the steadily increasing traffic load.
The traffic during the week is most intense at Strelsau International Airport.
At weekends there are peaks of arrivals at Hentzau and Zenda in the Morning, and peaks of departures in the Afternoons.
They also wished to evaluate DUMMKOPF as an aid for Strelsau International.
(This is an Artificial Intelligence based system developed by Professor V. Frankenstein of Strelsau University.)

lide 3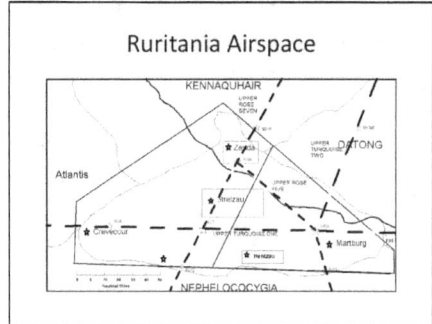

This shows the proposed division of Ruritanian Airspace, slightly modified from the original proposal to minimise coordination problems.
Unfortunately, with three airports, it is not possible to provide an equal division of workload between two sectors.
We compared the two sector arrangement shown with the current one-sector organisation.

Slide 4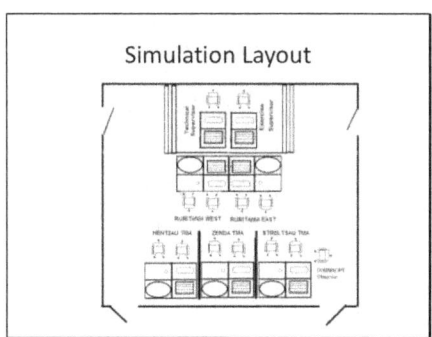

The simulation room was laid out with the two sectors side-by-side, with the technical and executive supervisors. The three TMA suites were placed opposite, with barriers to discourage direct communication – of which there was practically none.

The TMAs were not placed in accordance with their physical layout. Strelsau was placed at the end so that the boffin responsible for DUMMKOPF could look over without getting in the way.
In the one-sector organisation we used Ruritania West, giving a better view of Strelsau over the vacant Ruritania East suite.

Slide 5

Traffic Samples

- 95% ile traffic 2019
- Weekdays / Weekends
- Doubled to form data store
- Checked for anomalies
- Samples drawn at random for
- 50% / 80%/100%/125%/150%/175%
- No samples repeated.

We began by establishing the 95% peaks of traffic on weekdays and weekends.
As often occurs, there were distinct morning and evening peaks.
To generate traffic samples, we began by doubling the traffic, establishing modified call signs, and running the results through our checker
to ensure there were no conflicts on arriving in the airspace.
We tweaked the data to solve any problems we found at entry - but not within sectors.
When we required samples, we took random samples of as many flights as needed.
We never repeated any sample.
There were, of course, common flights in different samples,
but this corresponds better with reality than completely random samples.

Slide 6

Planned Simulation Running

- 3 Weeks – 4 days+ 1 reruns – 4 Runs/day
- 2 days training
- 2 days 100%
- 2 Days 125%
- 2 Days 150%
- 2 Days 175%
- 2 Days 100%

We planned to run two-day blocks at steadily increasing traffic,
with a final repeat of the original 100% density (not of course the same samples)
We planned on running two blocks per week with a fifth day to cover mis-runs and other failures.

Control Room Simulation

Slide 7

Simulation blocks

- 2 days = 8 x 90 min Runs
- 1 or 2 sectors
- Weekday or Weekend
- AM or PM

Each Simulation block consisted of eight 90 minute runs,
containing all combinations of 1 or 2 sectors, Weekdays or Weekends, Morning or afternoon.!
We actually used two teams of controllers, which were equally balanced.
We also arranged to try to balance runs with and without DUMMKOPF.

Slide 8

Actual Running

- ½ Day Familiarisation   2 x 50% Training
- 1 day 80% Training
- 2 days 100% (3 lost - rerun day 5)
- 2 Days 125% (1 lost – rerun day 5)
- 2 Days 150% (none lost)
- 2 days 175% (2 lost – 3 Overloaded)
- 2 Days 100% (none lost)

In practice, training went fairly well, but the first measured block lost 3 runs out of 8 –
mostly because of problems with DUMMKOPF!
The 125% and 150% blocks worked out well,
But the 175 % block was a catastrophe –
3 exercises were halted almost at the start, and 2 were lost for other reasons.

Slide 9

Exercises Analysed

- First 100% too many disturbed (Dummkopf)
- 125% OK - 150% OK
- 175% Overloaded – not completed
- Final 100% OK
- DUMMKOPF – Strelsau TMA
- Hentzau / Zenda TMAs

Finally, we were able to analyse three blocks of eight exercises – 125%, 150% and the last 100%.
The first 100 % was too disturbed by DUMMKOPF – induced Errors,
and the 175% block was clearly overloaded.
We also analysed the three TMAs separately, analysing Strelsau for the presence or absence of DUMMKOPF.

Slide 10

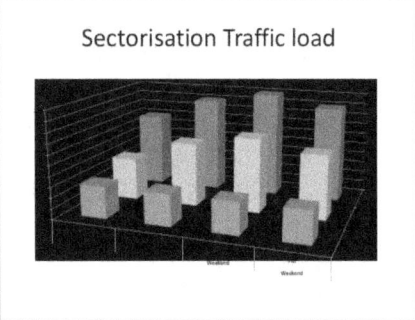

This figure shows how the traffic workload (in terms of aircraft present during the peak hour) divides between RUW (Ruritania West) and RUE (Ruritania East) , compared with the current single sector RU
This is the basic split of traffic.

Slide 11

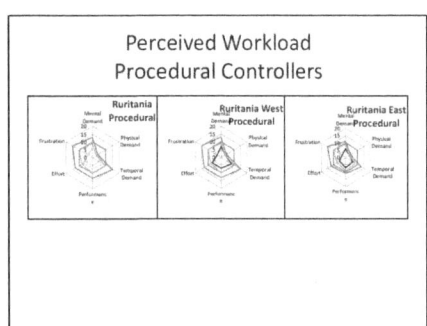

These Nightingale Diagrams show the average NASA TLX Component scores for procedural controllers
The regular hexagon corresponds to mid-scale levels, usually considered reasonable work levels.
The Physical demand level is practically negligible in all circumstances.
Although Mental demand is relatively low, Temporal Demand and Frustration are increasingly high with higher traffic levels
Ruritania West has virtually the same perceived levels as the single sector organisation.
Ruritania East is relatively underloaded, although fairly frustrated at higher traffic levels.

Slide 12

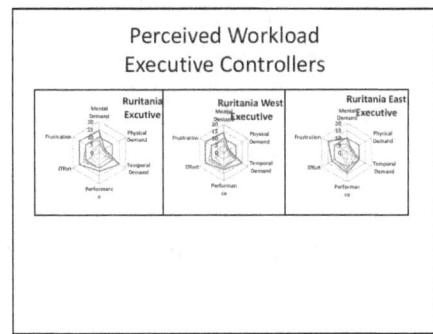

These Nightingale Diagrams show the average NASA TLX Component scores for executive controllers
As in the previous pcture, the regular hexagon corresponds to mid-scale levels, usually considered reasonable work levels.
Again, the Physical Demand level is practically negligible in all circumstances.
Although Mental demand is relatively low, Temporal Demand and Frustration are increasingly high with higher traffic levels
Ruritania West has virtually the same perceived levels as the single sector organisation.
Ruritania East is again underloaded, although fairly frustrated at higher traffic levels.

Slide 13

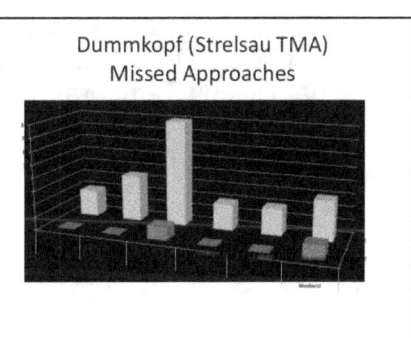

This shows the incidence of missed approaches in Strelsau TMA with and without DUMMKOPF.
The effect of DUMMKOPF is visible, increasing more than proportionately to the traffic, as the controllers lost the picture, and had to order go-arounds – not a normal procedure.

Slide 14

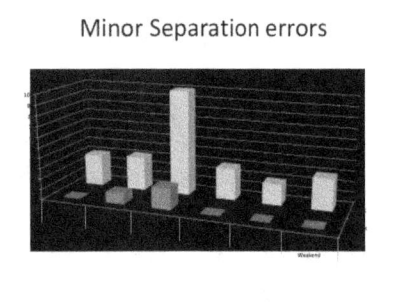

This shows the incidence of minor separation errors in Strelsau TMA with and without DUMMKOPF.
(A minor Separation error is a closest approach of 4-5 Nautical Miles without vertical separation)
There were no major separation errors – less than 4 Nautical Miles
The effect of DUMMKOPF is visible, increasing more than proportionately to the traffic, as the controllers lost the picture

Similar effects are visible in measures of delays in arrivals and departures.

Slide 15

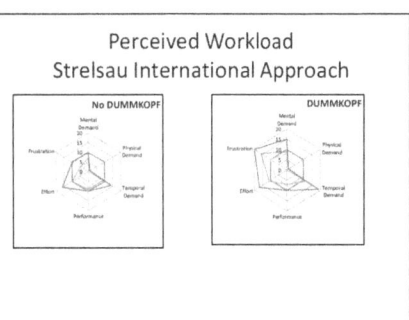

The controllers experience is reflected in these Nightingale diagrams
(There was no appreciable difference in morning and evening peaks.)
Although the negligible physical demand, performance and effort did not differ, Mental Demand, and particularly Temporal Demand and Frustration were greater when DUMMKOPF was in use.

Slide 16

Controller Opinion

"Where an aid is not trusted, the controllers are obliged to observe its recommendations, try to understand their reasons, evaluate them for safety, and maintain an on-going picture of what action to take in case of an error or malfunction. This considerably increases workload, time stress and frustration."

One controller expressed the effect of DUMMKOPF with some restraint like this.
Controllers were not hostile to the idea of an aid, but made a number of valuable suggestions which have been forwarded to the appropriate destination.

Slide 17

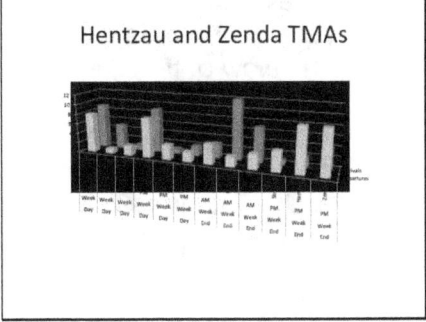

Hentzau and Zenda TMAs

Contrary to our expectations, it was not at Strelsau, but at Hentzau, and to a lesser extent Zenda, that the most urgent problems occurred.
In particular, weekend morning arrivals were higher than expected, as were weekend evening departures. Partly this involved a large number of training flights operated by the RCCATI (Ruritanian Civil and Commercial Aviation Training Institute) which operates from Hentzau. Some relief could be found by stopping training during peak periods.

Slide 18

Hentzau TMA - Overload

- About 25% of the arrivals and departures at Hentzau, on weekdays and weekends, consists of training flights operated by the Ruritanian Civil and Commercial Aviation Training Institute. These flights should be halted during peak periods.

In fact, it is unlikely that the RCCATI will expand as fast as the holiday traffic, and it is likely that it will be quite happy to avoid peak times, (Particularly if it is charged more at peak times.)

Slide 19

Conclusions
Weekday Operation

- Will the proposed division of Ruritanian Airspace provide increased capacity during normal week-day operation?

- *Yes, the proposed division into Ruritania East and Ruritania West will provide about 20% increase in week-day capacity, giving 170% of current capacity.*

A 20% increment would keep the system going for roughly another three years.
Bearing in mind that we are talking about overloads on about 20 days a year It might be simpler to impose some form of flow control on those days.

Slide 20

**Conclusions
Weekend Operation**

- Will the proposed division of Ruritanian airspace provide increased capacity during week-end operation?

- *Yes, the proposed division into Ruritania East and Ruritania West will provide about 20% increase in week-end capacity, giving about 160% of current capacity.*

Weekends are more stressed
because there is a tidal effect
An inflow of traffic in the mornings
An outflow in the evenings.
It may be possible to spread the traffic
using some form of flow control
so that peak traffic is avoided.

Slide 21

DUMMKOPF

- Is the Digital Universal Monitoring and Mediation Kontextual Operational Planning Function (DUMMKOPF) an acceptable function for Ruritanian Airspace?
- *No.*

Controllers judge a system as a whole.
Although they knew that there were teething troubles, They - quite understandably - could not make allowances.

Several controllers suggested that Real-Time simulation was not the way to test this sort of innovation, which requires evaluation for highly unlikely events.

Slide 22

DUMMKOPF PROBLEMS

- *There are certain situations where the DUMMKOPF function 'loses' aircraft, leading to potential conflicts..*
- *Controllers did not trust it. (see above!)*
- *There were problems in integration with the EEC simulator, which are likely to recur in its integration into the Ruritanian Airspace Control system.*

Slide 23

Further Conclusions

- The existing system can cope with up to 150% of current peak traffic during weekdays and up to 140% at weekends.
- The most loaded part of the system will be Hentzau TMA, not Strelsau
- The general level of equipment and procedures in Ruritanian ATC is obsolescent.
- A more modern system is required.

In reality, it is anybody's guess when the system will become overloaded. Just because traffic has grown steadily in the past does not guarantee it will in the future. Some economic catastrophe – such as Rurexit or a cheap synthetic coprolite could upset any predictions.

Slide 24

Thanks to :
Controllers from Ruritania Control
Controllers from Strelsau International, Hentzau and Zenda TMAs
Professor V Frankenstein + Dct I. Igorovitch of Strelsau University
Ruritania Air Traffic Control

Dct = Dozent = Ph. D. student, more or less.

Slide 25

Any Questions?

Chairman : Ask about the Pig?
EB - : I can't tell you about that here, but buy me a drink or two after the session, and I will TELL ALL.

Strictly NOT for official consumption.

Appendix 5 - Poster

The poster opposite is practically illegible at this scale

It would normally be at least four times this size, and use colour to emphasise the diagrams.

Note that Liz Bennet has cannibalised most of this from the report or conference paper.

It would be nice to do everything from scratch, but in the real world there is rarely the time or effort available for that.

Ruritania I Simulation
E. Bennet, EREWHONTROL Experimental Centre

Sectorisation

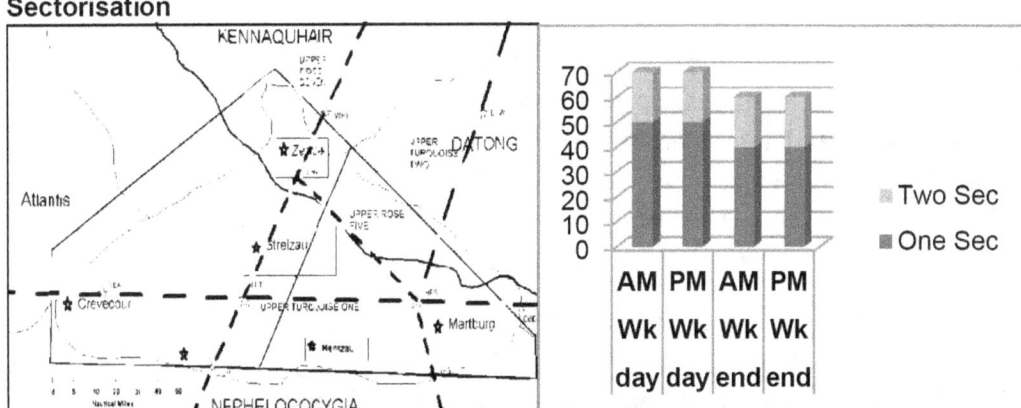

A two-sector organisation was introduced, and compared with the existing One-sector organisation.
Traffic used was contemporary 95%ile peak traffic, Weekend and weekday, morning and evening.
The existing organisation had an additional capacity of 50% for Weekday and 40% for weekend traffic.
There was no significant difference between morning and evening peaks.
The two-sector organisation added 20% capacity to these figures.

DUMMKOPF
(Digital Universal Monitoring and Mediation Kontextual Operational Planning Function

An AI-based function, DUMMKOPF was developed by Prof. V. Frankenstein of Strelsau University. It did not improve performance. In practice, increases were observed in missed approaches, (see diagram) minor separation infringements, arrival and departure delays Controllers stated *"Where an aid is not trusted, the controllers are obliged to observe its recommendations, try to understand the reasons, evaluate them for safety, and maintain an on-going picture of what action to take in case of an error or malfunction This considerably increases workload, time stress and frustration."*

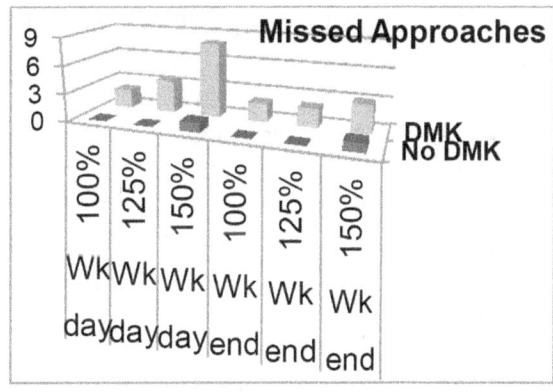

TMAs

Contrary to expectation, Hentzau TMA was the most heavily loaded, rather than Strelsau International.
Traffic was heaviest for Hentzau and Zenda Weekend Morning Arrivals,
Some relief can be given by rescheduling training flights and applying flow control at peak periods

About The Author

Hugh David became interested in computing in 1958, when studying Mathematics at Manchester University, where he had to register as an 'extra-mural' electrical engineering student to learn to program the Manchester Mercury computer, a valve machine. After National Service in the Royal Artillery, in 1961 he joined a design team developing a portable computer for field artillery. He designed and built a model of the interface with the user, using a 'window' system to prompt the user.

He then joined the Department of Ergonomics and Cybernetics at Loughborough University, producing an M. Sc thesis on the cybernetics of libraries. While teaching in that department, he completed a PH. D. thesis on the ability of air traffic controllers to predict potential conflicts in 1970.

From 1970 to 2002 he worked at the Eurocontrol Experimental Centre, studying Air Traffic Controllers and Air Traffic Control by methods ranging from neurophysiology to social anthropology.

He spent much of this time involved with the development, measurement, running and analysis of large-scale Real-Time simulations, acquiring practical experience and expertise not available in published form. He also developed a suite of statistical analysis programs (STATCAT) in ANSI FORTRAN IV, designed to provide a user-friendly interface with the user, before the term was coined. Elzevier published the complete package in 1982 as a 750-page book.

In 1996 he was created Independent Research Fellow at the Eurocontrol Experimental Centre, and carried out a series of studies on the measurement of strain and stress on controllers. He also began an examination of the displays used by controllers., which led to 'in-depth' studies of the Air Traffic system. .

He is continuing to study Air Traffic Control and controllers, as well as the complete Air Traffic system. He is currently completing a 'skeleton' re-design of the Air traffic system, applying systematic design to the complete system, to show that an efficient, economical, satisfying system can be provided by using existing technology in a planned fashion, rather than the present 'kludge' a collection of ad-hoc expedients